the earwig's tail

the
Earwig's Tail

a modern bestiary of multi-legged legends

May R. Berenbaum

harvard university press

cambridge, massachusetts london, england 2009

Library of Congress Cataloging-in-Publication Data
Berenbaum, M. (May)
  The earwig's tail : a modern bestiary of multi-legged
legends / May R. Berenbaum.
      p.   cm.
  Includes bibliographical references and index.
  ISBN 978-0-674-03540-9 (alk. paper)
  1. Insects—Popular works.   2. Arthropoda—Popular works.
3. Errors, Scientific.   4. Common fallacies.   I. Title.
  QL467.B46   2009
  595.7—dc22   2009013733

To the memory of the first scientist I ever met—my father,
Morris B. Berenbaum (December 19, 1924–October 11, 2006).
A shining example and role model of how to meet challenges
armed with knowledge and good humor, he was my hero
and I miss him.

the twenty-first-century insectiary

THROUGHOUT THE MIDDLE AGES, a form of literature called the bestiary was enormously popular. A bestiary is an illustrated compendium of "beasts," representing the animal (and occasionally plant) inhabitants of the natural world. More than simply scientific accounts of natural history, bestiaries were fundamentally religious works in that descriptions of beasts generally included some kind of moral or religious lesson; thus, any given bestiary is an illustration of the powerful cultural symbols of the particular era during which it appeared.

Modern, well-educated, and technologically sophisticated Americans look at bestiaries today and are amused by the willing suspension of skepticism that led the public to embrace unhesitatingly such improbabilities as mermaids, unicorns, griffons, dragons, barnacle goose trees, and manticores. In the medieval era, factual content and accuracy in animal stories were occasionally trumped by the religious allegorical significance or symbolism embodied in the stories. Remarkably, almost a millennium after the heyday of bestiaries, even among the most culturally literate populations in the most technologically advanced nation on the planet, there remains an extraordinary willingness to suspend skepticism with respect to wild stories about nature. One group of organisms, albeit a very large one, is particularly prone to misrepresentation—these are the insects and the rest of their jointed-legged relatives in the Phylum Arthropoda. Today's bestiaries—compilations of "facts" that fit people's worldviews despite the complete and utter lack of supporting scientific evidence—aren't painstakingly inked on vellum but rather texted and emailed through cyberspace.

After thirty years as an entomologist, I have come to realize that the majority of the most bandied-about insect facts familiar to the general public aren't facts at all. *The Earwig's Tail* describes my encounters with twenty-six of the most firmly entrenched modern mythical insects. In each, I track down the germ of truth that often inspires the misinformation and expand on the actual biology, which, by virtue of the amazing nature of the insect world, can be more fantastic than even the mythic misperception. Moral lessons are rarely explicitly stated in these accounts of arthropod life, but the tales are widely embraced and uncritically accepted at least in part because people often seek more than just natural history in nature stories.

I'm not saying there aren't life lessons to be learned from the study of insect biology. In general, though, I think it's probably not such a great idea to look toward insects for insight into the human condition or behavior to emulate—we're not really built for it, for one thing. It would be great to respect and admire them for the multitudinous ways they have devised for making their way in the world. These essays weren't written to expose and make fun of gullible people—far from it. The main purpose of the book is to highlight the strange and wonderful true-to-life details of insect biology and possibly, as a consequence, to encourage in the reader a healthy skepticism about what is factual and what is fictional about insects and their relatives. Perhaps such skepticism will even breed a willingness to believe the worst about them a little less reflexively.

One more note—many of the best-known bestiaries were written in Latin, the language of scholars of the day. Some of the Latin lingers in this modern bestiary, in the form of the scientific names of the arthropods in question. Probably the majority of nonscientists find Latin names off-putting, but they're a necessary evil because, with over a million species, there aren't enough common names to go around. The simple truth of the matter is that some insects just aren't plentiful enough to have acquired common names. Moreover, common names are notoriously imprecise. They vary from place to place—*Helicoverpa zea,* a caterpillar with a famously broad diet, is known variously throughout the United States as the cotton bollworm, the tomato fruitworm, and the corn earworm. The Latin name, though, is universally agreed upon, at least by entomologists. Common names have even given rise to some of the misperceptions people have about insects (the titular earwigs, for example, are rarely if ever actually found lurking in ears), so there's some value in using an ancient and less value-laden language.

Because a preface is supposed to provide explanatory remarks about a book, maybe there is one more thing I should mention about *The Earwig's Tail.* This book is a collection of stories that are intended to be funny—maybe not late-night talk show or standup comedy funny, but at least not in precisely the same mold as most science writing, which is often rapturously lyrical about the splendors of nature, passionately persuasive with respect to a particular crisis or controversy, or objective, impartial, and chockablock with facts. There's a very good reason why there aren't many people who write ostensibly funny science books: It's very hard to know what other people might find amusing. E. B. White, the brilliant writer of the arthropod-friendly *Charlotte's Web,* among many other books, said about the subject, "Analyzing humor is like dissecting a frog. Few people are interested and the frog dies of it." I haven't dissected a frog in decades, nor have I made a science of dissecting humor, but as an entomologist I have found insects to be an admittedly intermittent but nonetheless remarkably rich source of humor. In this book, I wrote about things that struck me as funny. I hope people are interested, and if it's any consolation, no frogs died in the process.

the beasts

the earwig's tail

the aerodynamically unsound bumble bee

THERE'S JUST SOMETHING fundamentally in-
compatible between the world of insects and the world of physics. Undoubtedly, the most dramatic example of that incompati-
bility is the old saw about bumble bees and aerodynamics. Search
the Web and you'll find that it's common wisdom that, according to the laws of physics, bumble bees can't fly. You can find the
story in all kinds of places, from self-help sites to science-bashing
sites, but all relate the same tale. As the story goes, at a dinner
party sometime during the 1930s (dinner parties seem to be a recurring theme in stories about scientists), a scientist was asked

about the flight of bumble bees and, after a few back-of-the-napkin calculations, he pronounced that bumble bees could not generate sufficient lift to get off the ground.

Subsequently, of course, it has been shown that bumble bees knew what they were doing all along; the apparent discord was a result of faulty assumptions on the part of the scientist. For one thing, insect wings are flexible; the initial calculations assumed that bumble bee wings are rigid and fixed in place, like the wings of an airplane. With wings that move, insects can generate lift in ways flying machines can't—some by a specialized fling mechanism, for example, whereby the wings move in a figure-eight pattern that moves air downward and backward, propelling them forward and upward. Bees have wings with a rigid leading edge and a flexible trailing edge. These create spinning masses of air, or vortices, that hold them up on the downstroke, creating what is known in the aeronautic world as dynamic stall. In addition, bees flap their wings furiously fast—over 200 beats per second —to compensate for the predicted decrease in aerodynamic performance associated with their small size. This not only permits them to stay airborne, it also allows them to tote around heavy loads of pollen and nectar. As a result, they make that buzzing sound for which they have become so famous.

I've heard the bees-can't-fly story many times and in many places and I've always assumed that it was a physicist with the napkin at the dinner party that eventful evening. Jacob Ackeret, a Swiss gas dynamicist, is often credited with the calculation, probably because he was well known in the field of supersonic aerodynamics at about the right time in history. The sad reality is that the source of the story was neither Swiss nor a gas dynamicist but rather was August Magnan, a French entomologist, who wrote an otherwise obscure scholarly text on insect flight titled

Le Vol des Insectes (Magnan 1934). On page 8, he wrote, "j'ai appliqué aux insectes les lois de la résistance de l'air, et je suis arrivé avec M. Sainte Lague à cette conclusion que leur vol est impossible." ("I have applied to insects the laws of air resistance and I have arrived with Mr. Sainte Lague at the conclusion that their flight is impossible.") How the concept entered popular culture without attribution is unclear, but it has been firmly ensconced in the popular conscience ever since.

So it was an entomologist and his colleague, the mysterious M. Sainte Lague, apparently some kind of laboratory assistant, who were responsible for this notion. It's reassuring to know that at least I'm not the only entomologist who has problems with physics. Although I've been interested in biology for as long as I can remember, I am forced to admit that, throughout my childhood, many other scientific disciplines failed to captivate me. In fact, for as long as I can remember, I've hated physics. Maybe "hate" is too strong a word, but I really have a visceral dislike for the subject. I consider this a personal failing, particularly in view of the fact that I feel obligated as a professional scientist to have an interest in all scientific subjects. I'm not certain how this aversion came about—I expect, however, that the college course I took in physics, taught by a very new assistant professor who seemed particularly vulnerable to flirtation by attractive female students (a group to whose ranks I did not belong), and the high-school course in the subject that I took, taught by the school's assistant principal, whose other charge was serving as school disciplinarian, undoubtedly did little to foster an interest.

A big part of my problem with physics was the way that it was presented. Physics problems were almost invariably deeply disturbing. I didn't keep my physics notes from high school, but a visit to contemporary Web pages featuring physics problem sets illustrates the point (and confirms that things haven't changed much): Physics problems are either scary or depressing. Consider a problem I found at the University of Oregon's Web site:

> A 1,400 kg car, heading north and moving at 35 miles per hour, collides in a perfectly inelastic collision with a 4,000 kg truck going east at 20 miles per hour.
> a. What is the speed and direction of the wrecked vehicles just after collision?
> b. What percentage of the total mechanical energy is lost from the collision?

And traffic isn't bad just in Oregon. Problems found at a University of California at Berkeley physics problem set Web page were similarly distressing, irrespective of the type of physics being illustrated by the problems. To illustrate the physics of acceleration,

> While driving down the highway at 30 m/s (which is about 72 miles per hour) John spots a police car, sirens flashing. At exactly 1:00 P.M., when the police car is also traveling at 30 m/s and is 100 meters behind John's car, John floors the gas pedal, giving his car an acceleration of 2 m/s^2. John keeps the pedal floored for 5 seconds. During this time, the police car continues moving at 30 m/s, and the officer radios for backup. How fast is John's car moving at the end of those 5 seconds?

As for momentum conservation laws:

> A bomb explodes into three fragments. Immediately after the explosion, the first fragment, of mass $m_1 = 2$ kg, travels leftward at $v_1 = 100$ m/s. The second fragment, of mass $m_2 = 3$

kg, moves at a 45° angle as shown, at $v_2 = 80$ m/s. The third fragment, of mass $m_3 = 4$ kg, moves at the angle shown, with velocity $v_3 = 50$ m/s. Was the bomb moving before it exploded? If so, what was its speed and direction?

For me, doing physics homework was too much like watching the evening news without the "feel good" stories—all bombs, explosions, and traffic accidents, unremitting examples of irresistible forces meeting immovable objects.

So imagine my surprise and delight when an undergraduate honors student, Nathan Van Houdnos, turned in a term paper in my general education entomology class entitled, "Word Problems Involving Insects as Educational Tools in Mathematics and Physics." Nathan pointed out that word problems involving insects are found fairly frequently in both mathematics and physics for similar reasons—insects are small, easily visualized, useful for three-dimensional parametric equations (due to their capacity for flight), and, for the vast majority of the public, disposable (so that being on the receiving end of collisions isn't too disturbing). At the University of Illinois, all mathematics majors take a course called "Fundamental Mathematics," which introduces students to techniques of mathematical proof. Midway through the semester, they encounter the classic (in certain circles) "fly and train" problem: "A runaway train is hurtling toward a brick wall at the speed of 100 miles per hour. When it is two miles from the wall, a fly begins to fly repeatedly between the train and the wall at the speed of 200 mph. Determine how far the fly travels before it is smashed" (D'Angelo and West 2000). All of my years of entomological training leave me ill-prepared to answer this question (other than to deny emphatically that any fly alive can achieve a speed of 200 miles per hour). According to Nathan, however, this problem is solved by using the convergence of infinite series, and a very small object meeting the train is needed to allow that approach to work. Insects have the size and mobility to make the problem at least seem plausible; a word problem conjuring up the image of a paramecium or a rotifer shuttling back and forth on train tracks might be too distracting to solve.

This fly-and-train problem is legendary among mathematicians in part because of its association with the great John von Neumann, the brilliant mathematician and innovative computer scientist. The story is that, when asked a version of this problem, he quickly provided the correct answer; when asked how he had obtained the correct answer so quickly, he replied, "Simple! I summed the series!" This is apparently a side-splitting anecdote among mathematicians.

Insects can, according to Nathan, be helpful in preparing students for applications of Newton's force law (the idea that the force on an object is proportional to its mass multiplied by its acceleration). From the sophomore physics textbook at the University of Illinois at Urbana-Champaign is this problem:

> A buzzing fly moves in a helical path given by the equation $\mathbf{r}(t) = \mathbf{i}b \sin \omega t + \mathbf{j}b \cos \omega t + \mathbf{k}ct^2$. Show that the magnitude of the acceleration of the fly is constant, provided b, ω, and c are constant.

Or, if your taste runs more to the order Hymenoptera,

> A bee goes out from its hive in a spiral path given in plane polar coordinates by $r = be^{kt} \theta = ct$ where b, k, and c are positive constants. Show that the angle between the velocity vector and the acceleration vector remains constant as the bee moves outward. (Fowles and Cassiday 1990)

Why a fly would traverse a helical path, or a bee a spiral path, is of little consequence to most physicists, I guess; I think the assumption of the word-problem writers must be that the behavior of insects is so bizarre that almost anything they'd be required to do to meet the assumptions of a physics problem would be possible, or at least imaginable in the minds of undergraduate physics majors, who probably feel about insects the way I felt about anything connected with physics when I was an undergraduate.

Insects are also useful in physics because they're so close to being point masses (i.e., mass is distributed at the center of an object, rather than over its height, width, and length), and to simplify things many problems have the student assume objects are point masses. So, what better way to study Newton's force law in rotating reference frames than with a cockroach on a turntable?

A cockroach crawls with constant speed in a circular path of radius b on a phonograph turntable rotating with constant angular speed w. The circular path is concentric with the center of the turntable. If the mass of the insect is m and the coefficient of static friction with the surface of the turntable is μs, how fast, relative to the turntable, can the cockroach crawl before it starts to slip if it goes *(a)* in the direction of rotation and *(b)* opposite to the direction of rotation? (Fowles and Cassiday 1990)

I suppose it doesn't matter, at least to physicists, that bees don't normally fly in a spiral and cockroaches likely wouldn't stay long on a moving turntable. But they shouldn't be too smug; insects have mastered some engineering feats that human engineers still can't explain. According to the calculations of another entomologist, Brian Hocking (made not on a napkin but in a 1957 issue of *Science Monthly*), those furiously flapping bees manage to get about 450 million miles per gallon of nectar; that kind of fuel efficiency is likely to elude the major airlines for the foreseeable future.

the brain-boring earwig

INASMUCH AS NEITHER is a discipline widely embraced by the general public, it's not surprising that many people confuse entomology, the study of insects, with etymology, the study of word origins. Occasionally, though, the two disciplines run at cross purposes. Take, for example, earwigs—a group of insects regarded as so peculiar even by entomologists that they are assigned their own taxonomic group, the order Dermaptera. These insects aren't particularly diverse—there are only about 1,800 species worldwide—and they tend to lead a rather low-profile existence. They are most often brown or black in color and rarely exceed about an inch (25 millimeters) in length (although the largest recorded species, topping out at 3.15 inches or 80 millimeters, is the giant earwig of St. Helena, *Labidura herculeana,* which may now be extinct). Earwigs do share one attribute to which they owe the ordinal name bestowed upon them by William Kirby in 1818. "Dermaptera," which means "skin wing," refers to the short, leathery front wings that characterize most members of the group. Most also have a long, flexible abdomen capped with a pair of pincers, called forceps. Earwigs use forceps variously for opening up their wings, grabbing mates during courtship, defending themselves, and immobilizing prey. There are a few exceptional species that are ectoparasites—that is, they live externally on the bodies of warm-blooded hosts—that have lost even these distinctive traits. About ten species live in the fur of giant rats in tropical Africa, eating what is euphemistically referred to as "scurf" (shredded skin flakes or scales), and another half-dozen species or so live on the bodies of bats in Malaysia; these oddballs are wingless and have forceps that are straight, rather than curved.

Their common name, however, is about as old as any name for an insect in the English language. "Earwig" derives from the Old English "ear *wicga,*" which, roughly translated, means "ear insect" or "ear wiggler" (*wicga* being the etymological basis for the word "wiggle"). This name supposedly reflects the venerable belief that earwigs have a predilection for crawling into people's ears and wreaking havoc—depending on sources, they may burrow into your brain or merely content themselves with laying eggs and hatching out a new brood of ear wigglers destined to drive insane their hapless host. The Oxford English Dictionary dates this etymology to the eleventh-century Saxon Leechdom, an early herbal. Its persistence over centuries is reflected by the virtual universality of common names for dermapterans. Nations that have agreed politically on no other issues seem to share the unshakeable conviction that earwigs are irresistibly drawn to ears. The French call them *perce-oreille* ("ear-piercer"), the Germans *Ohrwurm* ("ear-worm"), and the Russians *ukhovertka* ("ear-turner"); the same applies to Danish, Dutch, and Swedish. Even the great eighteenth-century systematist Carolus Linnaeus, who devised the two-part scientific naming system still in use today and who came up with names for over 2,000 insect species, made reference to the idea in naming the common European earwig *Forficula auricularia* (with *auricula* meaning "ear").

Like so much entomological misinformation, the notion that earwigs infest ears may have originated with Pliny the Elder, first-century polymath who, among other things, believed that caterpillars originate from dew on radish leaves. According to Philemon Holland's 1601 translation of his *Historia Naturalis* (Pliny's ambitious yet ultimately unsuccessful effort to catalogue all knowledge), "If an earwig . . . be gotten into the eare . . . spit into the same, and it will come forth anon." Not long after, Nicholas Culpepper provided an alternative method for extracting earwigs in his 1652 *The English physitian: or an astrologo-physical discourse of the vulgar herbs of this nation:* "[Hemp juice]. . . is held very good to kill the Worms in man or Beast, and the Juyce dropped into the Ears killeth Worms in them, and draweth forth Earwigs, or other living Creatures gotten into them."

In view of the fact that "hemp juyce" is derived from *Cannabis sativa,* the marijuana plant, I wonder if some of those claiming to have earwigs in their ears may have imbibed the stuff rather than dropped it into their ears. Although entomologists generally like to rationalize this persistent notion that earwigs like to crawl into ears by explaining that many earwigs, particularly the most commonly encountered ones, seek out moist, dark places, which aptly describes most auditory canals, I find it curious that I've only been able to find one single reference in about ten centuries of literature to an earwig actually being found in an ear, which hardly seems common enough to merit a common name. I'm not the first to notice this discrepancy; George William Lemon, in his 1783 *English etymology; or, A derivative dictionary of the English language,* was equally baffled by the putative origin: "[w]ig here seems to carry the idea of *wriggle,* or, as we sometimes say, *wiggle waggle;* and consequently *an earwig means the insect that wriggles itself into the ear;* though an instance of such an accident was perhaps never known; or, if ever it happened, must have happened so seldom, as scarce to have been sufficient to affix an appellation to this creature; we may therefore very much doubt even this deriv. and yet I am unable to produce a better."

This absence of an abundance of reports of earwigs in ears is not for lack of a literature of insects in ears; a veritable zoo's worth of arthropods has been reported over the centuries in ears of one sort or another. In more recent times, Ryan and colleagues (2006) reported that, according to unpublished data from the Johns Hopkins emergency department, the most common foreign objects in ears of adults were cockroaches; Bressler and Shelton (1993) also reported that cockroaches were the most common foreign objects in the ears of ninety-eight patients. Another review evaluating the insecticidal activity of reagents used to remove "insect foreign bodies of the ear" lists at least two species of cockroaches, honey bees, and beetles as "most frequently" requiring removal, along with at least one noninsect, a tick (Antonelli et al. 2001). Most memorably, O'Toole and colleagues (1985) related the case of an unfortunate patient who presented with a cockroach in each ear, affording the team of physicians an extraordinary opportunity to conduct a "controlled trial," comparing two different methods of removal from the same patient.

Thus, of all the arthropod fauna reportedly found in ears, earwigs are conspicuous by their absence. Cockroach invasions of aural cavities are understandable, given the tendency of cockroaches to infest houses. Many earwigs are also found in homes, but they're usually restricted to cracks and crevices in damp, musty basements, not kitchens, bedrooms, and other rooms in which people (and their ears) are most frequently found. Moreover, the reluctance of earwigs to fly would seem to reduce the probability of their gaining access to the ears of anyone who doesn't habitually sleep with his head jammed into basement corners.

After days of searching for even one example of an earwig in an ear, I was delighted to come across a report in the *Rocky Mountain Medical Journal* titled, "The earwig: The truth behind the myth" (Taylor 1978). I had to wait a few more days to read the paper because I had to order it through interlibrary loan, and when it finally arrived, I couldn't help feeling a little bit let down. Instead of a photograph of the specimen, there was a cartoonish drawing of an earwig. There was no description of how the specimen had been handled and identified, who had identified it, and how it might have gained entry.

There is at least one alternative etymological explanation for the connection between "ear" and "earwig," offered (without attribution) by Frank Cowan (1865). Although the front wings of earwigs are short and leathery, their hind wings, which fold up and tuck underneath the short front wings, bear an uncanny resemblance to a human ear in shape when unfolded. It could be that earwigs earned their moniker based on their morphology, and gradually etymology became destiny. Although entomologically this explanation is a little more satisfying, etymologically the evidence is not really compelling. That Lemon (1783), desperate as he was to find an alternative to the unsatisfying "ear wriggle," made no mention of it suggests that the explanation may be of relatively recent origin.

It's a shame that about the only thing people think they know about earwigs isn't generally true. Laying eggs in a place where their kids couldn't survive just doesn't fit the earwig profile; all known free-living earwigs display a remarkable degree of maternal care, keeping watch over their eggs and newly hatched nymphs, feeding them and protecting them from erstwhile predators. And the frightening-looking forceps, which have inspired most of the other common names of earwigs (including the common name "pincerbug" and the Latin *Forficula*, meaning "little shears"), aren't even the most bizarre anatomical feature; for reasons not exactly clear to entomological science, males of the earwig family Anisolabididae have a spare penis. Although the females have only one genital opening, male anisolabidids have a pair of organs, one of which points in what would seem to be the wrong direction. While its function is not known, it's believed that the extra intromittent organ can be mobilized if something untoward happens to the slender, elongate primary penis (Kamimura and Matsuo 2001). One wonders what common name dermapterans might have acquired had this anatomical feature attracted Pliny's attention back in the first century.

the california tongue cockroach

THERE ARE MANY frightening things that cross my computer desktop via the Internet. Most come by email: pernicious viruses with the capacity to cripple my computer, requests for letters of recommendation from students who took a class from me so long ago that I no longer have any clear idea who they are, notes from production editors wondering when overdue manuscripts will be arriving, and many more too horrible even to mention. But there is one kind of email message I often receive that doesn't frighten me—the kind that starts with,

"This is a true story!" or words to that effect. Almost invariably, such notes contain what is called an urban legend.

According to Jan Brunvand, acknowledged world authority on the subject, an urban legend is a realistic story "concerning recent events (or alleged events) with an ironic or supernatural twist" (Brunvand 1981). In that they are a "unique, unselfconscious reflection of major concerns of individuals in the societies in which the legends originate," it's not surprising that some of the most venerable urban legends involve arthropods, well known to be sources of concern to a broad cross-section of western society. Most people harbor no great love for arthropods of any description and are thus more than willing to believe the worst about them, making them ideal subjects for urban legends.

I frequently get inquiries by email about certain urban legends; I expect it's because people know I have an interest in newsworthy arthropod feats, however improbable they may be. Two such inquiries have crossed my desk in recent years. One came to me from my colleague Art Zangerl, who forwarded a message he received from his wife, who in turn forwarded a message from one of her colleagues. It began, as do so many urban legends,

> THIS IS REALLY GROSS!!!!!!!!!!!!! BUT TRUE!!!!!!!!!!!! Be prepared, this is AWESOME !!!!! If you lick your envelopes . . .
> You won't anymore! A woman was working in a post office in California, one day she licked the envelopes and postage stamps instead of using a sponge. That very day the lady cut her tongue on the envelope. A week later, she noticed an abnormal swelling of her tongue. She went to the doctor, and they found nothing wrong. Her tongue was not sore or anything. A couple of days later, her tongue started to swell more, and it began to get really sore, so sore, that she could

not eat. She went back to the hospital, and demanded something be done. The doctor took an X-ray of her tongue, and noticed a lump. He prepared her for minor surgery. When the doctor cut her tongue open, a live roach crawled out. There were roach eggs on the seal of the envelope. The egg was able to hatch inside of her tongue, because of her saliva. It was warm and moist . . . This is a true story reported on CNN!

This story immediately struck me as implausible. Among other things, the cockroaches most likely to be found in California post offices—the Oriental cockroach *Blatta orientalis,* the brown-banded cockroach *Supella longipalpis,* and the American cockroach *Periplaneta americana*—all lay a dozen or more eggs cemented together in rather sizable suitcase-like packages called oothecae, so how a single egg could detach and work its way into a cut on a tongue wasn't clear to me. The story didn't make sense even if the egg made its way into the tongue while still ensconced in the ootheca, or egg case; a cut large enough to accommodate oothecae of the species of cockroaches most likely to be living in California post offices would have to be up to a third of an inch long and an eighth of an inch deep. Given the relatively rich blood supply to the tongue, a cut of that size would be producing so much blood that the postal worker probably would have had to seek medical attention right away (although it's conceivable that the pain from such a cut was what made her oblivious to the large, dark brown, purse-shaped ootheca stuck to the envelope).

Even if the entomological elements of the story, by virtue of the incredible diversity of the insect world, proved true, there's still the disturbing fact that, in the story, the postal worker is licking envelopes and postage stamps. Usually, in post offices I've visited, it's the patrons that lick envelopes and postage stamps, generally while waiting in line; the postal workers are the ones who put up the little "Window Closed" signs just as you reach the head of the line.

A quick check with my spouse (who is an amateur collector of urban legends) confirmed this story was in fact a hoary example of the genre. An Internet search revealed several variants. One warns, "You'll never eat fast food again!" and relates the story of a girl who ate a chicken soft taco from a popular fast-food franchise and ended up with a swollen jaw. Differences were subtle— the cockroach eggs ended up in the salivary glands and not the tongue, and they were removed (along with "a couple of layers of her inner mouth," or cheek) before they hatched, so no little baby roaches had an opportunity to make an appearance. Another difference is that the putative source is not CNN but rather the "November 19" *New York Times,* which, according to the *Times'* Web site, has no insect-related stories at all, other than one on pesticide use in cities. Yet another difference is that the girl is purportedly suing the restaurant, whereas in the post-office story no lawsuit or Mexican food is mentioned.

A more venerable cockroach-related urban legend, however, has circulated for the better part of a century without the help of the Internet. This one manifested in a story in the *Jerusalem Post* published August 25, 1988. According to the story, a woman frightened by a cockroach she spotted hurled it into the toilet and "sprayed it with a whole can of insecticide," failing to flush it in her panic. Her husband, ignorant of these activities, then went to "use the toilet, [and] dropped in a smoldering cigarette," setting off the fumes and creating an explosion that burned his "sensitive parts." When the ambulance arrived, the attendants, carrying him out on a stretcher, laughed so hard upon hearing how he had incurred his injuries that they "dropped the stretcher . . . down the steps of his house, causing further injuries; these were specified as 'two broken ribs and a broken pelvis'" (*Jerusalem Post,* August 25, 1988).

This story was immediately picked up by two international wire services, and it was reported in newspapers around the world. In the United States, the story appeared in dozens of newspapers, from the *Boston Globe* to the *Seattle Times.* Few felt any compunction at having a laugh at the expense of another person's pain. Puns abounded—"Victim Is Butt of Bad Joke," the *Detroit Times* reported. The *San Diego Tribune* proclaimed, "Woman Bugged by Roach but Spouse Suffers," and the redoubtable *Weekly World News* declared, "Man Bowled Over by Exploding Potty." Students of urban legends, however, immediately recognized this story as an updated variant of a classic that goes back to the days of outhouses and privies. In its original form, flammable material is dropped down a hole in an outhouse and an unsuspecting victim answering a call of nature suffers the consequences of lighting a cigarette (the punch line traditionally has the victim wondering, "What the heck I et?"). The exploding privy story goes back well over a half-century, and insects may even have played a role at an early stage: In some versions of the story, the flammable liquid is dumped into the privy for the purpose of killing maggots. Just like indoor plumbing in the contemporary version of the story, the cockroach may be nothing more than a concession to modern times.

The exploding toilet story was quickly debunked and exposed for what it was. Rather than express mortification at having published an implausible story without checking sources, many newspapers simply mined the retraction for more heavy-handed humor, with such headlines as "Paper Finds Bug in Tale of a Roach," or "Roach Tale Turns Out to Be Crawling with Errors, So Paper Steps on It"; the *Richmond News Leader* ran the headline,

"Old Exploding Toilet Story Has News Faces Flushed." It seems a shame that people insist on making bad jokes based on human fears and insecurities. After all, cockroach infestations present a serious urban pest-management challenge, and *Blatta* control is important even outside the bathroom.

Cockroaches do occasionally make legitimate newspaper headlines. One such event was the publicity stunt carried out at Six Flags Great America in Gurnee, Illinois, in October, 2006, in connection with Halloween. In a promotion called, "A cockroach is your ticket to the front of the line," anyone standing in line willing to eat a live Madagascar hissing cockroach could cut to the head of the line. There certainly was publicity; in fact there was a public outcry from several corners. A subdued warning was issued by Lake County health officials alerting the public that there might be health consequences to eating cockroaches of unknown provenance. And there was an impassioned outcry from People for the Ethical Treatment of Animals, although an online petition circulated to protest the stunt garnered fewer than 600 signatures by the appointed date. Let's face it, this benighted enterprise created little furor because the live animals being consumed were cockroaches. Rules of humane slaughter don't (in the United States, at least) apply to any living creature unfortunate enough to lack a spine. That they can be eaten alive actually seems to drive up the price of oysters.

Even as a vegetarian, I can understand why eating oysters that are alive doesn't present a moral dilemma to most people. As it happens, the behavioral repertoire of the legless, headless, eyeless live oyster isn't substantially different from the behavioral repertoire of a dead oyster. But the Madagascar hissing cockroach, known to entomologists as *Gromphadorhina portentosa*, isn't just any cockroach (as suggested by the specific name *porten-*

tosa, Latin for "marvelous" or "prodigious"). This species has complex social hierarchies, an elaborate communication system, and maternal behavior that includes provisioning of bodily fluids to offspring. Their behavior in many respects may be more sophisticated than that of the yahoos that dreamed up this promotional stunt. Theme-park representatives, oblivious to the nuances of cockroach life, countered the protests in press interviews by asserting that cockroaches were farm raised and fat free to boot. I found both claims amusing. Their claim that cockroaches are fat free doesn't square with the fact that entomologists refer to the organ that fills up most of their abdominal cavity as the "fat body." As for "farm raised," although the phrase conjures up images of contented cockroaches grazing in verdant pastures, I find it much more likely that those cockroaches were raised the way our department raises them—in large green pasture-free plastic garbage cans.

Apparently, one goal of the promotion was to take a run at the world record for eating live Madagascar hissing cockroaches— thirty-six in one minute, set in 2001 by Ken Edwards of Derbyshire, England. A check of Internet sources revealed that as of March 2007 Edwards still held the record, so I'm guessing the Gurnee event was a washout. But even had the record been broken, it would hardly have had much of an impact on the history of eating live arthropods. Eating human body lice, for example, has a long and storied history, running the gamut from Budi nomads and their descendant Kirghiz and Kazaks in eastern Europe, the Cheyenne and Snake Indians of North America, the forest-dwelling Mois in Cambodia, the Hottentots in Africa, and aboriginal people from a range of islands in Southeast Asia and Oceania. Although such practices have fallen into abeyance throughout western Europe, there remain two pockets of entrenched live entomophagy. In Germany eating animals while they're still alive is technically verboten, but prosecution is less than vigorous with respect to Milbenkäse, or, as it's sometimes known, Spinnenkäse—mite or spider cheese. In Würzburg, this variant on German Altenburger cheese is considered a great delicacy. The cheese is deliberately infested with *Tyroglyphus casei,* the cheese mite. According to my translation from the German of A. Hase (1929), "Millions of this species live in hard cheese, ultimately transforming it into a grey, mobile powder made up of the mites, their cast skins, and their excrement." Lest you think this is just lip service, the extent to which this regional dish is beloved is demonstrated by the existence of a cheese mite monument in Würchwitz.

Perhaps even less appetizing than cheese with mobile mites is cheese with mobile maggots, a Sardinian delicacy known as *casu marzu* (literally translated as "rotten cheese"). This product is prepared by allowing Pecorino Sardo, a sheep-milk cheese popular in the region, to become infested with the larvae of *Piophila casei,* the cheese skipper. Over time, the cheese becomes soft and weepy (exuding a liquid known as *lagrima,* or "tears" in Sardinia but known to forensic entomologists as the "black putrefaction" stage of decomposition). The dish has been vividly described as "a viscous pungent goo that burns the tongue and can affect other parts of the body" (Trofimov 2000). The maggots can actually survive passage through the intestinal tract, causing nausea, bloody diarrhea, and vomiting while in residence. These unfortunate side effects are the principal reason that it is technically illegal to sell *casu marzu,* but a thriving black market (or, one supposes, black putrefaction market) exists nonetheless (Overstreet 2003).

Given the side effects, Six Flags might want to consider importing *casu marzu* for its next promotional stunt. There's a catchy potential slogan—"a cheese skipper can be a ticket to skipping to the front of the line!" And there's also the added benefit that, if you manage to survive the roller-coaster ride without vomiting, you can look forward to the experience in the comfort of your own home for days afterward.

the domesticated crab louse

HAVING BEEN A CARD-CARRYING entomologist for almost thirty years (or at least a card-possessing entomologist, because I'm not exactly sure where my Entomological Society of America membership card is), I've seen a lot of insect species firsthand. Over the years, though, there have been conspicuous gaps in my experience. For example, it was over twenty years after I first took an insect taxonomy class that I saw my first live *Pthirus pubis*.

It's not that it's such a rare species; *Pthirus pubis,* also known as the crab louse or the pubic louse, is an ectoparasite that infests humans around the world (specifically around the groin region). Having seen photographs, drawings, and even preserved specimens, I knew what they looked like, so when I first encountered one I knew immediately what I was looking at. The circumstances, however, were a little awkward. People often stop by our entomology department's office if they have insect identification questions, generally bringing in little plastic bags or pillboxes or envelopes containing dead specimens in various stages of decomposition. In this case, a gentleman walked in concerned that he might have contracted Lyme disease because he had picked up some ticks. I happened to be standing next to him when he lifted up his shirt and pointed to one of the ticks, which was crawling along his waistline as he spoke; from my vantage point I could see instantly that the tick was in fact a crab louse. Not knowing exactly how to break the news to him, I asked a male graduate student who was also in the office to take the man aside and explain the situation to him while I tried to look preoccupied with weighty department matters.

As insect ectoparasites go, crab lice aren't the worst of the lot; unlike fleas, mosquitoes, and even their close relative the body louse, *Pediculus humanus humanus,* crab lice haven't been implicated in the transmission of any diseases. Mostly, they spend their entire lives nestled among hairs on the body of their human host, preferably in the pubic region, sipping blood about four or five times a day from the day they hatch to the day they die, a period encompassing about a month and a half. Their bites can cause intense itching and frenzied scratching can lead to infections and ulcerating sores. A crab louse can move to a new host only when its host comes in close contact with another human; given the neighborhood in which they reside, that close contact generally involves intermingling of body hair. Thus, like *Chlamydia,* syphilis, and gonorrhea, crab lice are most frequently transmitted via sexual intercourse.

Even though crab lice don't transmit any fatal diseases, most people aren't too happy about being infested, what with the intense itching, suppurating crab sores, and social embarrassment that can accompany an infestation. At least that's what I thought until I saw a post on the Listserv Entomo-L with the subject line, "Is it illegal to sell crab lice?" The subject of the query was a Web site, LoveBugz.net, billed as the "FanSite of the Lousing Lifestyle." This site purported to sell "specially bred pubic crab louses [sic] from Japan (not the same as homeless people's variety of lice exactly). First, they DON'T BITE, they just live off dead skin cells and such. . . Really, you're cleaner with them there than without them. Second, these babies are HUGE!! . . . And they just live happily in your underwear. It's so COOL! They grow, and have families. . . It's like having personal Sea monkeys in your pants."

Much of the ensuing discussion on the Listserv, and on the Internet in general, centered on whether or not this site is parody or paraphilia. There is a relatively rare form of sexual orientation, a form of zoophilia called formicophilia, in which people gain sexual satisfaction from intimate contact with ants. Etymology (and for that matter entomology) notwithstanding, the term "formicophilia" (from the Latin *formica,* for "ant," and *philia,* for "love") can also be used in connection with sexual stimulation caused by the crawling on or nibbling of the genitalia by small animals other than insects, including frogs or snails (Dewaraja and Money 1986; Dewaraja 1987). But the "lousing lifestyle" didn't seem to be about sexual gratification, at least as far as the lice are concerned. Part of the appeal, according to the Web site manager, Dr. Bugger, is in giving them to other people ("when you give them to someone else, it's like they become part of your family since their lovelice are the babies from mine").

Actually, the lousing lifestyle didn't exactly seem to be about pubic lice, either, or at least about pubic lice as they're known to entomologists. Amidst references to "crabs," "nice lice," and "love lice" were occasional references to "bed bugs" and "chinches," which are names reserved not for lice but for two species of bloodsucking true bugs. In fact, if, as claimed, the "love lice" don't consume blood, then they're not pubic lice, or any other kind of sucking lice, either, since all sucking lice require blood (not dead skin cells) to live.

I'm inclined to think that the site is a joke, but the fact that it's not immediately discernible is a sad reflection of how little really is known about one of about a half-dozen insects that can't live without us. There are only two species in the genus *Pthirus;* the only living relative that the pubic louse has is *Pthirus gorillae,* the gorilla louse. Molecular evidence suggests that about 3 million years ago some errant or adventurous gorilla lice ended up infesting humans, giving rise to the lineage leading to today's *Pthirus pubis.* This conclusion was based on molecular data analyzed by David Reed and his colleagues at the Florida Museum of Natural History. These authors speculated that the transfer came about as a consequence of human consumption of gorilla meat, or possibly human use of abandoned gorilla nests, sidestepping the delicate question as to whether zoophilia is deeply rooted in the evolutionary history of *Homo sapiens.* For that matter, although Reed and his colleagues described in great detail how their genetic analyses were conducted (2007), they were brief in their description of how they collected those gorilla lice in the first place, and I can't help but wondering, given that *P. gorillae* occupies habitat comparable to that preferred by the crab louse, who among the authors was the lucky one assigned that task.

However humans were first infested, we've had a long association with *P. pubis*. Crab louse remains have been found in material collected from a Roman pit in England dating back to the late first century AD (Kenward 1999) and from 2,000-year-old mummies found in the Atacama Desert in Peru and Chile (Rick et al. 2002). But this close relationship may be coming to an end. Armstrong and Wilson (2006) analyzed the annual incidence of sexually transmitted diseases, including *Chlamydia*, gonorrhea, and pubic lice at the Department of Genitourinary Medicine at Leeds between 1997 and 2003 and found that, while gonorrhea and *Chlamydia* increased dramatically, the prevalence of pubic lice dropped significantly. These authors note that the study period coincided with the rise in popularity of a procedure known as the "Brazilian," a cosmetic pubic hair removal procedure resembling a bikini wax except that virtually no hair is left behind. Although the association is correlative, it's sobering to contemplate. Extensive depilation is gaining popularity among men as well as among women. Habitat loss is a major factor contributing to species extinctions across the planet; the same phenomenon may be happening in our pants. Arguing for the preservation of a human ectoparasite that prefers to infest our private parts will be an interesting exercise, but I sure don't want to be the one who writes the habitat conservation plan.

the extinction-prevention bee

JUST ABOUT EVERYONE who knows anything about insects is familiar with the phenomenal communication abilities of the honey bee, *Apis mellifera*. With their famed "waggle dance," honey bee foragers can convey precise and complex information about distance, location, and abundance of floral resources in the kind of symbolic language once thought to be exclusively the province of *Homo sapiens*. As well, this little insect, with a brain only one-millionth the size of a human brain, can use subtle chemical signals to convey information with amazing rapidity to thousands of nestmates about threats to the colony, status of the food supply, and viability of the queen.

So it's more than a little ironic that the ability of people to communicate on the subject of honey bees is curiously deficient. There's been a lot of talk about honey bees of late, due to the sudden, devastating appearance of what has come to be known as colony collapse disorder—the mysterious rapid decline of colonies, leaving just a handful of workers tending an apparently healthy queen along with brood and food stores (van Engelsdorp et al. 2007). Honey bees, of course, are the nation's premier managed pollinator and are responsible for commercial pollination of close to a hundred crop plants. This enigmatic disappearance, with its enormous implications for the American food supply, has proved to be irresistibly attractive to the media.

Of course, the principal attraction is the opportunity to work a pun (which for want of a better word must be called bee-labored) into a headline. The *Dallas Morning News* remarked that the "Strange disorder has scientists, beekeepers buzzing" (April 24, 2007) while the *New Haven Register* more succinctly summarized the situation with the headline "Buzz, off" (April 30, 2007). The *Washington Post* declared "The flight of the honeybee: A mystery that matters" (May 9, 2007), the *Boston Herald* bee-moaned the fact that "Colony collapse disorder bee-devils farmers" (April 18, 2007), and the *Detroit Free Press* deemed colony collapse disorder "A sticky situation" (May 23, 2007). Meanwhile, the *Boise* [Idaho] *Weekly* was "Bee-fuddled" (May 23, 2007), the *Black Hills* [South Dakota] *Pioneer* "Bee-wildered" (May 7, 2007) and the *Springfield* [Illinois] *State Register* was "Feeling the sting" (May 19, 2007). Somewhat prematurely, perhaps anxious to work an alternative pun into the story, *Newsday* declared, "Experts may have found what's bugging the bees" (April 26, 2007). Most creative, I think, was the historically resonant headline, "The lost colonies," which appeared in the *Zanesville* [Ohio] *Times Recorder* (May 21, 2007), and the subtle yet apt Beatles (Bee-tles?) reference in the headline, "Give bees a chance," appearing in the online commentary magazine *The Simon* (May 1, 2007).

Accompanying almost all of the inevitable puns in the various and sundry headlines were dire warnings of the consequences of bee disappearances. A story in the influential German newspaper *Der Süddeutsche Zeitung,* Germany's largest national daily paper with a circulation over 600,000, provided a pithy assessment of the gravity of the situation from the undisputed scientific genius Albert Einstein: "Wenn die Biene von der Erde verschwindet, dann hat der Mensch nur noch 4 Jahre zu leben," or, loosely translated, "If bees disappear from the earth, humans will cease to exist within four years." I came across this story not because I'm in the habit of perusing German periodicals but rather because I was interviewed for the story and the journalist sent me a copy. I was quoted in the story as saying, among other things, "Wenn Sie einen Hamburger essen . . . dann verdanken Sie das indirekt den Bienen," which is, roughly translated, "Whenever you eat a hamburger, you have a bee indirectly to thank." I'm sure my high-school German teacher would have been pleased by the grammatical correctness, but, as pithy or quotable phrases go, it certainly falls far short of the Einstein quotation, in either language.

As for that Einstein quotation, it certainly sounded authoritative and credible, particularly in German. Even in translation, however, it didn't sound familiar. I'm no expert on Einstein, but over the past three decades, I've made a practice of collecting pop-culture references to insects. (Did you know, for example, that Johnny Depp had an insect collection? Or that Henry Fonda was a beekeeper?) So I was surprised to have somehow missed this quotation altogether. Moreover, on reflection, I couldn't imagine in what context Einstein might have made the remark. Why would anyone ask a physicist, even the world's most famous physicist, about bees, and what circumstances would prompt him to offer his views on pollination to the world? I searched through what I could find of his writings and located only a single reference to bees, which appeared, reasonably enough, in a 1949 essay on socialism:

> It is evident, therefore, that the dependence of the individual upon society is a fact of nature which cannot be abolished— just as in the case of ants and bees. However, while the whole life process of ants and bees is fixed down to the smallest detail by rigid, hereditary instincts, the social pattern and interrelationships of human beings are very variable and susceptible to change. Memory, the capacity to make new combinations, the gift of oral communication have made possible developments among human beings which are not dictated by biological necessities. (Einstein 1949)

Although the passage mentioned bees and people, there was not even a fleeting reference to any utility of bees beyond their metaphorical significance. Einstein's apocalyptic yet apparently apocryphal quotation, though, was quickly making the media rounds. In print it appeared on both sides of the Atlantic—in the *Independent* and the *Telegraph* in the United Kingdom, in the *International Herald Tribune,* and in dozens of small American papers and Internet blogs. The comedian Bill Maher even mentioned it during his show *Real Time* on HBO on April 20, 2007. It didn't really bother me to see the quotation in all kinds of places, but when it showed up in a grant proposal on which I was a co-principal investigator, I thought I should probably check it out.

As it turns out, I was certainly not the only one who couldn't find the quotation in any of Einstein's writings. Before I stumbled across it, the Web site Snopes.com, devoted to quashing Internet rumors, had already dispensed with questions surrounding its authenticity (April 21, 2007), reporting that at least one Einstein biographer, Walter Isaacson, and the author of *The New Quotable Einstein,* Alice Calaprice, had never come across it in their extensive research. According to the site, the quotation appears not to have existed before 1994, almost a half-century after Einstein died. So, if Einstein did indeed say it, he must have said it at a séance through a medium.

The quotation appeared to have materialized for the first time in a pamphlet published by the National Union of French Apiculture in the midst of concerns throughout Europe about unfair price competition from cheap honey imports and looming tariff reductions predicted to exacerbate the problem. In the pamphlet, beekeepers warned of the dire consequences of a collapse of their industry, invoking Einstein in predicting that honey dumping by China could well mean the end of human civilization on earth.

This is the story according to Snopes.com. Although I couldn't find any 1994 newspaper stories online containing the quotation, I'm inclined to believe it because it's consistent with the general crisis in world beekeeping that resulted from Chinese honey-dumping around that time. Beekeepers are hardly the first to resort to fabricating quotations—they may not even have been the first to fabricate one by Einstein (Snopes.com reports that he is also purported to have said, "compounding interest is the most powerful force in the universe" in 1983, only twenty-seven years after his death). It's a familiar ploy—to give a concept credibility, what better strategy than to fabricate a quotation from a revered authority? It doesn't even matter that Einstein worked in a field totally unrelated to beekeeping—he is probably the scientist most familiar to the American public, despite the fact that he's been dead for over fifty years. In fact, it's almost a mark of honor to be the source of an invented quotation; it suggests some measure of respect for one's reputation, if not for publishing standards and practices.

Now, I've been misquoted in the past—the most egregious example occurred several years ago, when a newspaper quoted me using the phrase "spiders and other insects." I am positive that not even massive quantities of mind-altering drugs or weeks of sleep deprivation could reduce me to a state in which such words would pass my lips. Even elementary-school students know that no insect has more than six legs whereas all spiders, which are arachnids, have eight legs. Spiders and insects are placed in different classes, just as rabbits (Class Mammalia) and robins (Class Aves) are; I would never use a phrase like "rabbits and other birds," either. Journalists have quoted me splitting my infinitives, dangling my participles, and referring to "data" as if the word were a singular, rather than plural, noun. But colony collapse disorder was a personal milestone for me in that for the first time I was the source of a fabricated quotation. On April 23, 2007, I was searching Google News to see whether there were actually any new developments in colony collapse disorder, and I saw my name in a story from a source, a site called "Smooth Operator," with which I was fairly certain I'd never had contact. I clicked on the link and read:

> The phenomenon of disappearing bees was first noticed late last year in the United States, where honeybees are used to pollinate $15 billion worth of fruits, nuts and other crops annually.

Scanning down the article to find my name, I saw:

"The main hypotheses [sic] is that Kevin Federline is stealing
the bees," said May Berenbaum, an insect ecologist at the University of Illinois, Urbana-Champaign. Berenbaum has never
liked K-Fed and blames him for turning former wife and pop-
sensation Britney Spears into the laughing stock of the entertainment world . . . In some cases, beekeepers are losing 50
percent of their bees, with some suffering even higher losses.
One beekeeper alone lost 40,000 bees. Nationally, some 27
states have reported the disappearances. In each instance, the
bee disappearances coincided with a K-Fed concert, book
signing, or paternity suit.

Needless to say, I never said it—although frankly it's as plausible as many of the other hypotheses that have been proposed in
cyberspace (including, but not limited to, cell phones, changes in
the Earth's magnetic field, thinning of the ozone layer, global
warming, genetically modified crops, cannibal bees, automobile grilles, honey bee "rapture," Chernobyl, wireless Internet,
Osama Bin Laden, and alien abduction). As I've written to the
dozens of people who have sent me email containing their hypotheses to explain colony collapse disorder, while we can't definitively rule out their hypotheses, they are inconsistent with
what is known about the disorder, particularly its epidemiological distribution.

But back to the central question—would mankind survive to
see its next leap year if bees disappeared? As annoying as I find the
term "mankind" (inasmuch as the planetary majority of *Homo
sapiens* lacks a Y chromosome), it could indeed survive without
honey bees. Among other things, the vast bulk of calories in-
gested worldwide—mostly from wheat, rice, corn, or other

grains—are contributed by plants that don't need any pollinators at all. And although bees do pollinate the majority of fruits, nuts, and vegetables, fortunately for the future of humanity many other sources of fruits and vegetables rely on pollinators other than bees. Onions and cacao (the source of chocolate) are pollinated by flies, figs are pollinated by wasps, and several tropical fruits, including durian, are pollinated by bats. So, although our diet may be considerably duller, at least we wouldn't be entirely bee-reft of fruits (or puns, for that matter).

the filter-lens fly

IN JUST ABOUT EVERY insect fear film ever made, there's an obligatory insect-eye view of a potential victim. There's a general recognition on the part of filmmakers that insects possess compound eyes with many facets and the way this anatomical feature is rendered in movies is through use of a multi-image filter lens, which, depending on film budgets, repeats an identical image tens or dozens or hundreds of times. In *Empire of the Ants*, director Bert I. Gordon has his giant ants, created by exposure to toxic, radioactive waste, eyeing dozens of Joan Collinses in as many wet, clingy blouses to great effect.

In reality, what insects actually see wouldn't make for a very scary (or titillating) scene in a movie. As far as entomologists can determine, the insect compound eye produces a mosaic sort of image, like the image created by thousands of dark and light dots in a black-and-white newspaper photograph. Although no one is absolutely certain, the general belief is that insect eyes can't create images with high resolution, but that the compound eye is exceedingly good at detecting motion. So those giant ants in *Empire of the Ants* probably didn't have a very clear picture of Joan Collins, but they could probably see with relative ease the heave of her bosom as she screamed.

Heaving bosoms aside, insects and movies have a long history of association, dating back even to the earliest days of cinema. A fly, for example, is said to have inspired the invention of animation around the turn of the twentieth century by Segundo de Chomón, a Spanish filmmaker. Filming intertitles a few frames at a time for a silent movie, the filmmaker noticed the apparent jerky movements of a fly inadvertently included in the footage and realized that repositioning an object between each frame of film creates the illusion of motion when the film is played back at normal speed. Long though it may be, however, the association has never been an easy one, particularly when it comes to pressing insects into service as actors. In the early twentieth century, would-be documentary filmmaker Wladislaw Starewicz discovered, in his attempts to film the territorial battles of stag beetles, that when powerful stage lights (needed to provide sufficient illumination for the cameras) were turned on, the beetles stopped all semblance of normal behavior. The resourceful Starewicz realized that the beetles were much more easily manipulated once they were dead and painstakingly wired appendages back onto beetle carcasses and positioned them to his liking while reanimating them on film.

Not all filmmakers of the era, though, relied on such draconian measures to capture insects on film. Almost forgotten today are the pioneering efforts of one Mr. F. Percy Smith, who, according to a chronicler of the time (Talbot 1912) possessed "the happy faculty of investing his subjects with a quaint fascination which compels appreciation." Mr. Smith wanted to produce a film that would illustrate to his audience "the physical energy possessed by the common house-fly." Smith relied on conditioning to get his flies to behave the way he wanted them to. A fly was imprisoned in a dark box equipped with a thin glass door at one end; this door had a small opening into which was fitted a toothed wheel that rotated freely. According to Talbot (1912),

> the imprisoned fly, seeing the daylight entering through the glazed end of the box, attempted to escape in that direction, but found its passage obstructed by the glass. When it struck the latter, it received a smart tap on the head from a tooth in the wheel, which was caused to move through the fly's frantic efforts. Time after time the fly threw itself against the glass door, and on every occasion it received a rap on the head. At last frenzy gave way to tractability, and it came to the conclusion that the best means of escape was by walking up the wheel. Of course, as it advanced the wheel slipped round in the opposite direction. While the insect was walking like a criminal on a treadmill, the pictures were taken.

Smith modified his approach somewhat to film flies outside the box, tethered in place, and in this way was able to obtain footage of them seemingly juggling dumbbells, corks, bits of vegetables, other flies, and sundry other objects. When the film was released "the newspapers far and wide associated the cinematographer with strange powers, and the capacity to train the bluebottle in much the same way as a lion tamer subdues the King of the Forest."

Percy's effort clearly demonstrated that a fresh and inventive mind is needed to work with insect performers; after all, it's unlikely that repeated raps on the head to encourage obedience was a technique used with human actors of the day. Today, there is a legion of inventive Hollywood "insect wranglers" whose job it is to manipulate the behavior of insects on cue using whatever means available. Perhaps best known is Steve Kutcher, who has provided his insect-wrangling services to dozens of Hollywood productions. He includes in his arsenal for motivating arthropod actors such tools as "hot air guns . . . , vibrating wires (they don't cross over them), and Lemon Pledge furniture wax," which not only insures that they hit their marks but presumably helps to keep them shiny and scratch-resistant, too (Loud 1990).

This all inevitably leads to one question—can insect actors actually watch themselves once the films are released? This question assumes they live long enough, an unlikely prospect for two reasons. For one thing, postproduction editing can take a year or more, pushing the limits of lifespan for many species. This biological limitation is further constrained by the treatment that used to be accorded to arthropod actors under U.S. quarantine laws. The entire set of the feature film *Creepshow*, for example, was sealed after the segment titled, "They're Creeping Up on You" wrapped and all 20,000 Trinidadian cockroaches specially imported for the project were gassed (which likely wouldn't have happened had they thought to join the Screen Actors Guild prior to filming). But assuming they could live long enough to see their movies and that they had enough spare money for a movie ticket, the question remains—could they actually watch the movie?

Humans see movies as a result of what's called flicker fusion. If a light is flashed on and off at low frequency, the human eye perceives these individual flashes. If the frequency is increased, although individual flashes are not perceived, there remains an awareness of "flicker." At some point, however, the flicker effect disappears and all that is perceived is a steady light. For humans, the flicker fusion frequency is on the order of forty-five to sixty flashes per second in bright light and twenty-four to thirty flashes per second in dim light. This is why films today are projected at twenty-four frames per second with three-bladed projectors. Each of the blades covers the image three times per second while the film advances through the projector, so the end result is seventy-two screen images per second.

The flicker fusion frequency (FFF) has been measured for a wide range of species, generally with electrodes connected to the retina or light-sensing surface and with exposure to light of a particular intensity. Intensity matters because of the Ferris-Porter law, whereby FFF varies according to the log of illuminance (the amount of light hitting a surface). Whether insects can watch movies would thus seem to depend on species; grasshoppers, with a FFF around thirty-five, might enjoy watching themselves drown in *Beginning of the End,* whereas blow flies, with a FFF on the order of two hundred, might find viewing *The Fly* an experience comparable to paging through a family photo album one picture at a time.

It's difficult to guess at what the Asian swallowtail butterfly *Papilio xuthus* might see at the movies. The compound eye of the Asian swallowtail is equipped with photoreceptors of five different spectral types, with peak sensitivities around 600 (red), 520 (green), 460 (blue), 400 (violet), and 360 (ultraviolet) nanometers. Each of these different photoreceptors apparently has a different FFF, with the green receptor having the highest maximum at 107 hertz and the violet receptor having the lowest maximum at 82 hertz (Nakagawa and Eguchi 1994). Maybe this is why one typically doesn't see butterflies in line to buy tickets for the latest Technicolor extravaganza; it must be disconcerting to see Elizabeth Taylor's violet eyes gaze fluidly while the green grass she's walking on waves gently frame by frame.

But, according to Cole Gilbert, things aren't even that simple. Dr. Gilbert is an insect physiologist who specializes in insect vision. Several years ago, during a telephone conference call that was part of an external review of the entomology programs at Cornell University, I had the occasion to ask him whether anyone knew if insects could watch movies (okay, I admit it—I'm easily distractible). He thoughtfully provided me with a reply by letter several days later. In his letter, he explained that higher-order processing complicates things tremendously, citing work by Franceschini (1985). The explanation, which is greatly abbreviated here, actually didn't clarify things much for me:

> He [Franceschini] looked at the electrophysiological response of a well-known motion-sensitive, directionally selective, visual interneuron in blow flies while stimulating with a high tech "theater marquee" . . . The bottom line is that, even with photoreceptors and lamina monopolar cells fast enough to resolve changes in illuminance out to 200 Hz, the flies may still see movies as continuous motion due to slower higher-order processing in their motion-sensing pathway.

So the long and short of it is that I still don't know what insects can see when they go to the movies. It shouldn't bother me, but it does—I like to try to see the world as insects do. But I guess I should reconcile myself to the idea that some things may be beyond the human capacity to imagine. It's not just what insects see at the movies that has me wondering, either. I am quite amazed by the Asian swallowtail, the aforementioned *Papilio xuthus*. Its compound eyes, with their five different flicker fusions, are remarkable enough, but Arikawa et al. (1997) reported that these butterflies also have two pairs of ultraviolet/violet-sensitive photoreceptors on their genitalia. One pair is connected to motor neurons that control external genitalia, so that light stimulation "induces local movements of the genitalia, such as abdomen-curling, penis-withdrawal, and valve-opening." Arikawa and his colleagues suggest that these photoreceptors play a role in providing information to the male about the positioning of female genitalia during copulation. That may well be. But if I find myself sitting next to a male Asian swallowtail in a dark movie theater, I'm changing my seat before the feature starts.

the genetically modified frankenbug

THE BIOTECHNOLOGICAL TRIUMPH of genetic engineering, born out of the genetics revolution of the twentieth century, has provided benefits never before dreamed of—human growth hormone for people with growth disorders, for example, brewed up by a vat of bacteria rather than extracted from the glands of human corpses. Unfortunately, the concept has also given people nightmares. Much of the world has an unshakeable conviction that genes should stay in the genomes they arrived in and not insert themselves in any other organism's dance card. The concern is over unexpected consequences—a modern-day version of the sci-fi movie cliché that there are bounds beyond which science shouldn't go.

Thus it's not surprising that insects figure prominently in stories about genetic engineering experiments gone horribly awry; insects and unintended consequences are long-time partners in popular culture (as fifty years of big-bug horror films attest). The infamous lovebugs of the southeastern United States, known to entomologists as *Plecia nearctica,* are subject of persistent rumors that their existence is the result of human tampering with insect genomes in the name of biological control. I suppose this misconception is not such a stretch; they are a little peculiar looking. These small black flies with a red thorax and a tiny head that looks like it's mostly eyes are almost always seen flying around *in copulo.* This distraction accounts in part for the frequency with which they end up as dead couples on windshields and automobile grilles. In Internet lore, this fly is the result of an experiment designed to create a dipteran femme fatale that could mate with male mosquitoes and produce no offspring, presumably stealing his sperm so that none would remain for female mosquitoes. As the story goes, a male bug was created and a pair escaped to reproduce in near-plague proportions, tormenting the population of Florida on an annual basis by depositing their tiny bodies on car hoods, windshields, windows, walls, and fenceposts. Scientists at the University of Florida were blamed for this particular exercise in pushing science beyond all decent bounds.

Actually, University of Florida scientists weren't even the ones who first discovered *Plecia nearctica;* that distinction belongs to the Hawaiian fly expert D. E. Hardy, who named the species in an article in the *Journal of the Kansas Entomological Society* (Hardy 1940). And lovebugs aren't even native to Florida; they didn't arrive there until 1949, probably half-dead on automobile grilles, and didn't become particularly abundant for another two decades. In reality, the lovebugs are just fairly ordinary-looking flies in the family Bibionidae. Their larval lives are completely unremarkable—they live in soil and consume dead vegetation. Their adult life is difficult to ignore, however. After emerging as adults from the soil, males form large mating aggregations; females emerge a few days later and fly into these swarms, where they're likely to be grabbed by males. The pair heads for nearby vegetation and mates. Although sperm transfer is complete after about twelve hours, the pair remains coupled for days. The fact that females orient to heat, vibration, shiny surfaces, and certain chemicals that resemble those in automobile exhaust is why so many of them end up playing in traffic with fatal results (Hetrick 1970).

Genetic engineers are far more likely to move genes from one organism to another than they are to create entirely new insect life forms, a practice that nonetheless raises a whole new level of concern. I encountered a manifestation of this concern while in a local restaurant waiting in line to order lunch. A heated discussion was in progress between the counterperson and a customer, a female undergraduate. She was loudly declaring that, as a vegetarian, she didn't want any fish genes in her tomatoes. A quick glance at the menu on the blackboard confirmed my suspicion that she was speaking generically and not customizing a sandwich order, so I assumed that she was expressing her concern about genetically modified organisms (GMOs). I was a little embarrassed to realize that, despite being a vegetarian for over thirty years, I hadn't really given this particular aspect of genetic engineering much thought (doubly embarrassing because I'd been eating tofu hot dogs since before the undergraduate was old enough even to say the word "tofu"). I'd heard about fish genes in tomatoes. Specifically, genes encoding antifreeze proteins from Arctic flounder were engineered into tomato plants in order to determine whether these proteins might protect the fruits against tissue damage caused by the formation of ice crystals upon freezing. Although the antifreeze proteins were expressed, the fruits produced by the transgenic tomato weren't appreciably less damaged by freezing than were wild-type tomatoes. That's where, apparently, the project ended for DNA Plant Technology of Oakland, California, the company conducting the study. The frost-tolerant tomato with fish genes lives on, however, in hundreds of Web sites decrying transgenic technology (although sometimes it's a strawberry with fish genes).

What matters more about the exchange in the sandwich shop is that it got me thinking as a vegetarian entomologist about GMOs. It was a point of pride of sorts that the first animal gene to be expressed in a plant came from an arthropod; this was the luciferase gene of *Photinus pyralis,* a firefly, which in 1986 was successfully engineered into tobacco plants. Luciferase is the enzyme that, in combination with the substrate compound luciferin, the cellular fuel ATP, and oxygen, makes fireflies glow, an activity they engage in to find each other in the dark for mating purposes. In tobacco plants, luciferase makes tobacco foliage glow. The function of the glow is just to alert scientists that a particular genetic construct is working. This experiment did more than give rise to ponderous jokes about "lighting up"; it marked the beginning of a great adventure for firefly genes. Since that time, luciferase has been expressed in petunias, wheat, corn, and even tomatoes. Transgenic glowing tomatoes are not yet commercially available, despite the appeal of featuring them under the heading "light lunch recommendations" on menus everywhere.

As a Jewish vegetarian entomologist, I have to think as well about the implications of insect genes in odd places in the context of the laws of Kashrut, the kosher laws laid out in the Old

Testament. Leviticus 11:22 is explicit on the subject—the only kosher insects are those "that go upon all fours which have legs above their feet, wherewith to leap upon the earth." This means that saltatorial (jumping) types such as locusts and their kin are fair game. There's no mention of fireflies in the relevant passages, however. So the question arises—if a nonkosher gene is spliced into a kosher animal, does it make the kosher animal nonkosher? This question touches on both law and practice. In terms of practice, the "one in sixty" exemption has been applied. Halakha (Jewish custom) allows any nonkosher contaminant or additive as long as it constitutes less than "one part in sixty" of the new mix. Interpretation varies, however, as to the nature of this exemption. If the criterion is that the transgene constitutes less than one part in sixty of the genome (or if the protein product is less than one-sixtieth of the protein content), then arthropod genes in tomatoes or other vegetables (or in kosher animals such as cows) are probably all right. But others (e.g., Steven Druker, executive director of the Alliance for Biointegrity) have argued that the one in sixty exemption doesn't apply if that which is added has a "perceptible effect." It would be hard to argue that a salad that glows in the dark is not perceptible.

In terms of law at issue is *VaYikra,* or sundering the species boundary. Leviticus 19:19 states in essence that domestic animals cannot be crossbred and a single field cannot be sown with two types of seed; "thou shalt not let thy cattle gender with a diverse kind; thou shalt not sow thy field with two kinds of seed; neither shall there come upon thee a garment of two kinds of stuff mingled together." I'm not particularly well-versed in Jewish dietary law, but it would seem that this passage is a tough one to get around in terms of GMOs; then again, it would seem that I have to rethink those cotton/polyester-blend t-shirts in the closet as well.

Fortunately, not many arthropod genes are making the rounds of kosher animals. The luciferase gene has been expressed in glass catfish (to produce glow-in-the-dark fish for people bored with neon tetras) and in brine shrimp, but, since neither catfish nor shrimp are kosher, the presence of an arthropod gene doesn't affect their status much. About the only other arthropod gene that has been expressed in anything other than an arthropod is the sericin gene, the gene that encodes the major protein constituent of silk in spiders. In 1999, investigators at Nexia Biotechnologies in Montreal noticed similarities between the milk-producing glands of ruminants and the silk-producing glands in spiders. They managed to produce transgenic African goats (one not surprisingly named Webster and another less explicably named Peter) that carry the sericin gene in their genome. Bad luck for Nexia that both of the kids were male, given that male goats don't produce milk—but the transgenic goats were immediately put to work siring the next generation of nanny goats, which ought to produce silk protein in their milk (perhaps giving rise to a new generation of "Got silk?" commercials).

In terms of genetic engineering, turnabout is fair play. Although insect genomes have donated a few of their genes to other organisms, they've been on the receiving end as well. Given the ethical and logistical problems that arise when cloning vertebrates, it's not surprising that investigators have found it much easier to express vertebrate genes in insects rather than the other way around. The fruit fly *Drosophila melanogaster* is generally the organism of choice. Expressing vertebrate genes in invertebrates doesn't seem to have the shock value to the general public that expressing animal genes in plants has had; in fact, the same flounder antifreeze protein that caused such a fuss in tomatoes was expressed in fruit flies years ago without much fanfare. In a now classic paper, Halder and colleagues (1995) demonstrated that the eyeless gene of *D. melanogaster* encodes a transcription factor that controls eye development. A transcription factor is a protein that can turn other genes on and off. These authors inserted this gene into different fly tissues and managed to induce eye structures on legs, wings, and antennae. In and of itself, that feat (to say nothing of those legs, wings, and antennae) was impressive, but the investigators didn't stop there. Noticing sequence similarity with the mouse gene Small eye *(Pax-6)*, they incorporated mouse *Pax-6* cDNA in place of the *ey* cDNA and managed to induce eyes on a fly leg. So at one point in time mouse DNA was directing eye development in a fruit fly. It's likely that cDNA from the human gene *Aniridia* would have the same effect, and it may well be just a matter of time before *Aniridia,* a human gene, finds itself somewhere in an insect genome.

Splicing human genes into insects poses certain ethical problems, but insect genes will probably never be cloned into the human genome; most people hold insects in sufficient contempt that the idea of exploiting insect genes would be a nonstarter, irrespective of whatever remarkable products they encode. It's a shame, in a way—we're missing opportunities to become literally wasp-waisted or beetle-browed. You could put a bug in someone's ear for real. I get butterflies in my stomach just thinking about the possibilities.

the headless cockroach

As a movie-going scientist, I like to watch out for the science in science-fiction films. I find that usually this doesn't detract much from the experience, because scientific explanations rarely occupy more than a couple of sentences in the script. Particularly in insect-related science-fiction films, it's usually a simple matter to identify the entomological errors in the statements made by the supposedly knowledgeable characters in the film—as, for example, when Dr. Gates, Dr. Susan Tyler's entomologist mentor in the film *Mimic* (1997), expounds on the "kind of simplicity that governs the Phylum Insecta." Even first-year entomology students know that insects belong to the Class Insecta and the Phylum Arthropoda, two entirely different hierarchical categories of classification. But I have to confess that I was taken aback by a plot development in the sequel to *Mimic*, *Mimic 2: Hardshell*, which presented a challenge to my entomological expertise.

Mimic, described by critic Roger Ebert as "a loyal occupant of its genre," recounts how genetically engineered cockroaches in the subways of New York evolve to resemble humans and prey on them; in *Mimic 2*, the so-called "Judas breed" cockroaches are back and they seem to be selectively killing only those men that date Remi Panos, the entomologist schoolteacher who had only about ten minutes of screen time in the first film. Setting aside the myriad other improbabilities associated with human-sized man-eating cockroaches as well as the real-life relationship problems of female entomologists, the plot device that caused me greatest consternation was the plan concocted by Detective Klaski to dispatch the Judas breed cockroach by decapitating it with a paper cutter. Fortunately for single men in New York City, Remi knows that "you can't even make a scratch unless it's in molt . . . Besides, you know what happens when you take the head off a cockroach? . . . It dies, about nine days from now, when it finally starves to death." To make a long story short, the Judas breed is eventually decapitated with a pair of oversize scissors, leaving Remi and a surviving male character (who undoubtedly owes his good fortune to the fact that he's too young to be her boyfriend) trapped in her apartment while the dying Judas breed blocks the exit for eight more days.

It's always a little annoying to feel inferior to movie entomologists, so it was disconcerting to realize that I couldn't belittle the office supply–based integrated pest management plan that had been implemented because I didn't know how long cockroaches really can live without their heads. A search through the available literature reveals a remarkable dearth of longevity studies on decapitated cockroaches. This is not to say there is a dearth of studies on decapitated cockroaches. In fact, it's alarming how many studies there are that involve decapitation in general; a search of the 1993–2004 *Current Contents* database with the keyword "decapitated" brings up a disturbing 761. (Perhaps even more disturbing is the fact that only twelve of those studies involve cockroaches, but that's another story.) Taxonomically, decapitated cockroaches run a broad gamut; the death's head roach *Blaberus craniifer* (Goudey-Perriere et al. 2004), the discoid roach *Blaberus discoidalis,* the German cockroach *Blattella germanica,* the Madeira cockroach *Leucophaea maderae,* the speckled cockroach *Nauphoeta cinerea,* and the American cockroach *Periplaneta americana* have all been beheaded in the name of entomological science.

Perhaps the first scientifically motivated cockroach decapitation was conducted by G. A. Horridge of the University of St. Andrews in Scotland in what is now regarded as a classic study with the memorable title, "Learning of Leg Position by Headless Insects." Working with the American cockroach, Horridge (1962) demonstrated that

> a headless insect can be held in such a way that the legs receive small, regularly repeated electric shocks for as long as they fall into and make contact with a conducting saline solution. The legs, or as many of them as are not amputated, initially make many movements, some of which bring them into contact with the liquid surface, where they receive an electric shock . . . A commonly observed movement is a slow fall to the water surface and then, on receiving a shock, a sudden withdrawal or raising of the leg . . . Over a period of minutes

> . . . the legs are, on the average, raised with a greater effort
> and for progressively longer periods. Consequently less
> shocks are received.

So in essence, headless cockroaches are capable of learning to avoid shocks (as long as investigators leave at least one leg unamputated to have the opportunity to display learning behavior). For that matter, in subsequent studies aimed apparently at determining the absolute limit on how many cockroach parts can be removed by an investigator, Eisenstein and Cohen (1965) demonstrated that even isolated nerve bundles from the thorax, or midsection, of cockroaches are capable of avoidance learning.

Decapitation has allowed investigators to explore a wide range of physiological phenomena, including juvenile hormone biosynthesis, sex pheromone production, egg maturation and growth, and, somewhat more surprisingly, even a wide range of behavioral phenomena, including scratch reflexes, shock avoidance, and escape behavior. In fact, Ridgel et al. (2003) documented that aging cockroaches experience both locomotory and cognitive deficits and that, at least in aged *Periplaneta americana,* escape behavior is enhanced by decapitation (although these authors wisely refrain from speculating on the applicability of their findings to aged human subjects). In none of these studies, however, were headless cockroaches kept alive beyond the demands of the experimental bioassay period.

From where, then, has the widespread notion that headless cockroaches are viable arisen? An Internet search shows that the conviction that headless cockroaches of unspecified taxonomic identity can in fact survive for some period of time is persistent —but just how long is up for debate. An apparent majority of sites specify nine days, although there are sites that specify one week, two weeks, or several weeks. The sites differ as well in accounting for the ultimate cause of death in decapitated cockroaches, with about an even split between "starvation" and "dying of thirst."

Unfortunately, virtually none of these sites seems to be the least bit scientific. In fact, most have names like "Useless facts! Weird Information," "Totally Useless Trivia," or "Funky, Funny, and Freaky Facts." At least three sites claim that decapitated cockroaches live for exactly twenty-seven days, which is curious because it's well known, at least among fans of "Weird Al" Yankovic, that twenty-seven is a funny number, which would suggest that statements about decapitated cockroach longevity are perhaps being embellished for freaky, funky, funny effect. Consistent with this suggestion is the fact that, in addition to information on headless cockroach longevity, these sites report other biological observations that might be difficult to document, including: "Catfish have over 27,000 taste buds" (27,000 being almost as funny a number as 27), "A cat has 32 muscles in each ear," and "Dogs and humans are the only species that have prostate glands," among numerous others. A disproportionate number of these facts, by the way, involve sex and mating practices, and not all of them can be tastefully cited here.

It's depressing that seeking the answer to a legitimate entomological question is accorded the same importance by certain segments of society who create Web sites as, say, knowing the full name of the Skipper on *Gilligan's Island* (Jonas Grumby) or the state responsible for growing two-thirds of the world's eggplant (New Jersey). I suppose I should be grateful, though, that at least some screenwriters in Hollywood recognize that such biological information could be useful in some contexts. Somehow I don't think dropping Jonas Grumby's name would have helped Remi out of her dating crisis. As for me, despite the general absence of shocking negative reinforcement in *Mimic 2,* I've learned to look at office supplies in a whole new way. I am working on a novel approach to urban pest management using nothing more than a hole punch, a bottle of corrective fluid, and a staple remover.

I the iraqi camel spider

IT's BAD ENOUGH that U.S. servicemen in Iraq have had to deal with unbearably hot weather, improvised explosive devices, and the constant threat of land mines; it seems really unfair to throw giant camel spiders into the mix. In April 2004, the Internet was abuzz with photos of soldiers in desert camouflage holding aloft an oddly proportioned arachnid, its most distinguishing feature being that it appeared to exceed three feet in length. Text accompanying the photo claimed the creatures owe their name to their predilection for leaping several feet straight up into the air in order to latch onto a camel's stomach, whereupon they methodically suck its blood and lay eggs. They're also supposed to travel at speeds of twenty-five miles an hour and scream as they traverse the desert sand. Their ability to inject a potent anesthetic means that they can tear chunks out of a sleeping soldier, who won't know he has become a meal until he awakens to find he is missing critical bits of flesh. For what it's worth, they're also camel-colored (Walker 2004).

In reality, the Iraqi camel spider of Internet fame is a product of a combination of photographic foreshortening and arthropod eccentricity. The creature in the photograph, only about five or six inches in length before foreshortening, is in fact known in some professional circles as a camel spider. It's not, technically speaking, a spider at all, although it does belong to the Class Arachnida along with spiders and other eight-legged arthropods and thus shares with spiders eight legs instead of the six possessed by insects and two major body divisions instead of the three that insects have. Among its other common names is sun-scorpion, although it's not really a scorpion, either, and it and others of its ilk generally avoid, rather than seek out, the sun. The scientific name of the order, Solifugae, or "flee the sun" in Latin, is a more accurate representation of their behavior. Although the Iraqi version gets all the press, there are over a thousand species of sun-scorpions in the world, generally scattered among the desert regions of the world.

Sun-scorpions share carnivorous dietary habits with spiders, but they're unlike spiders in that they aren't venomous and they don't spin webs. Lacking venom to anesthetize or webs to snare, sun-scorpions rely on brute force to secure prey. They are equipped with two oversized jaws, or chelicerae, adjacent to which is a pair of long, leglike mouthparts called pedipalps, equipped with sticky tips. The pedipalps are used to find prey, which is then secured, sliced, and diced by the toothed chelicerae. Like spiders, sun-scorpions can't digest solid food and inject enzymes that liquefy the prey so that it can be sucked up.

Even though they don't use their powerful jaws to disembowel camels, they can use them defensively and have been known to inflict ragged, painful bites on unlucky humans. Most biting activity is confined to prey species, consisting of small fellow arthropods and the occasional lizard, so they don't deserve the various and sundry bloodthirsty nicknames they have acquired throughout the world. Known as camel spiders in Iraq, they are called *matevenados*, or deer killers, in Mexico. The notion that they seek out and chase after soldiers may be an accident of common interest; in seeking shade, they are likely to end up in the same places shade-seeking soldiers hang out. And reports of their speed are greatly exaggerated; top speeds are more on the order of ten miles per hour, although ten miles per hour is pretty impressive for an eight-legged animal and may be why they're sometimes called wind-scorpions.

This is not to say there are no arthropod dangers faced by American troops in the Gulf. Big threats come in small packages. In 1991, during Operation Desert Storm in Iraq, soldiers were beleaguered by attacks not only from Saddam Hussein's troops but by a myriad of biting arthropods, including an assortment of sand flies and midges often described as "sand fleas." A Rand survey conducted in 2000 in an effort to link pesticide exposure to Gulf War illnesses in veterans revealed that 3 percent of soldiers surveyed reported that "they had used flea and tick collars to protect themselves against insects during the Gulf War deployment." Collars were sent mostly by friends and families stateside eager to help out the troops.

The problem with this strategy is that flea collars are formulated for use on furry cats and dogs, not humans. Not surprisingly, given the toxic nature of the mostly organophosphate insecticides in use at the time, which were potent neurotoxins, a significant proportion, 5 percent, experienced adverse side effects. By September 1990, the Army Health Services Command issued a bulletin advising against the use of flea collars for any reason other than to protect dogs against fleas. Among other things, exposure to the neurotoxic agents in the collars could potentially affect the ability of a soldier to recover from exposure to enemy nerve gas. Because the practice continued through the war, another warning was issued in February 1991 by the Army's Office of the Surgeon General reiterating the message.

When troops returned to the Gulf for Operation Iraqi Freedom, the flea collars returned with them, necessitating the National Military Family Health Association to post another notice asking families more emphatically to refrain from interfering with military pest management efforts:

> Once again, well-meaning, generous Americans are thinking about servicemembers serving away from home and are looking at ways to help. Sadly, one of these methods promotes collecting pet flea and tick collars that will be added to "care packages." Once again, we're asking for your help to get word out to stop the practice.
>
> Most recently, we were made aware of a service organization's effort to gather the collars. In addition to pet flea and tick collars, the organization was encouraging local farmers to donate ear tags normally used with horses and cattle. While the idea to use the ear tags is new and appears to be limited to that particular local area, we continue [to] see media reports encouraging the sending of pet flea and tick collars . . . Last year during the early days of Operation Iraqi Freedom, the Armed Forces Pest Management Board alerted the Central Command Surgeon General of similar public ac-

tivities and advised that wearing the flea and tick collars is a dangerous practice which is harmful to the wearer and in violation of established federal laws.

In case the stern words were insufficient discouragement, the Armed Forces Pest Management Board (AFPMB) provided a graphic illustration of the adverse consequences of violating federal law by using flea and tick collars in a manner inconsistent with labeling.

Scrolling through the AFPMB's image library of sand fly bites in Iraq and Afghanistan, along with the parasitic infections by various and sundry microbes that can result, made me feel like a complete wimp for whining from time to time about a few measly flea bites. Living with four cats means an omnipresent risk of fleas and a few years ago I fell victim. There are times when I'd rather not advertise the fact that I'm an entomologist—it's not that I'm embarrassed by the profession, I'm afraid the profession will be embarrassed by me. During the summer of 2005, it seemed that I was inexplicably plagued by mosquito bites, most of which surprisingly seemed to occur indoors. Mosquito bites during the summer in central Illinois are hardly unusual, but the preponderance of indoor bites wasn't anything I'd experienced before. I was too distracted worrying about the probability of contracting West Nile fever, the mosquito-borne viral disease raging throughout the state at the time, and watching for symptoms to dwell on the significance of the fact that I wasn't actually seeing many mosquitoes either indoors or outdoors. I continued to worry about West Nile up until the moment I took one of our four cats, Splinter, to the veterinarian for a routine checkup. Dr. Spoerer noticed what she thought was flea dirt (the euphemistic name for the excrement of adult fleas) scattered around on the examination table underneath Splinter and suggested, reasonably enough, that Splinter might have fleas. I patiently explained to her that such a thing wasn't possible. Our cats lead a strictly indoor existence, never come into contact with other flea-carrying species, and weren't conspicuously scratching. In the late 1980s, I continued to explain, with a different set of cats, I used to have horrific flea problems, so I was well acquainted with how fleas operate, and there was just no way that our cats today could possibly have fleas.

While I was busy explaining all of this, Dr. Spoerer reached for a flea comb and began stroking it through Splinter's fur. After only about four strokes (while I was still recounting the statistical improbability of a flea infestation), she held up the comb, affixed to which was a small, flailing, flea-shaped, flea-colored insect that, with the benefit of thirty years of experience as an entomologist, I recognized immediately as *Ctenocephalides felis,* the common cat flea. It is an understatement to say I was mortified. Seeing the flea instantly clarified everything—the mysterious clusters of bites on my ankles, for example, where central Illinois mosquitoes rarely feed but where the cosmopolitan *C. felis* loves to attack. It explained as well the unbearable itchiness—it has been my experience as a host of a variety of insect parasites that flea bites are by far the itchiest and most persistent. At that moment, I also realized that I had wasted an inordinate amount of time watching for West Nile symptoms when I should have been watching for symptoms of murine typhus, cat scratch fever, tapeworms, bubonic plague, and other fleaborne problems.

Dr. Spoerer reassured me that great strides had been made in flea pest management since the late 1980s. I of course knew this already, on one level at least, inasmuch as I read entomological journals and attend entomological meetings; since my last personal experience with flea infestation, integrated flea management had moved away from treating the premises to spectacularly successful host-targeted therapy, with new, less-toxic products such as avermectins, Fipronil, juvenile hormone analogues (including pyriproxyfen), chitin synthesis inhibitors (such as lufenuron), and neonicotinoids. Dr. Spoerer prescribed a product that has been around for about a decade—Advantage, a formulation of imidacloprid, a neonicotinoid that targets nicotinic acetylcholine receptors in insect nervous systems. A one-time topical application of imidacloprid can reduce flea population on cats by over 95 percent for up to a month. Most of the fleas on the animal actually die within the first twelve to twenty-four hours.

I brought home enough Advantage for all four of our cats and applied it immediately. And the treatment lived up to its billing; the fleas did abandon the cats in droves. Soon the cats were effectively flea-free. Unfortunately, there remained unseen hordes of newly emerged ravenous fleas who now found the cats distasteful and who all managed to locate an alternative warm-blooded host. What a week earlier had been a few bites on my ankles became an all-out full-body assault. Oddly, neither my husband nor daughter seemed to attract any significant numbers of displaced fleas. In fact, the only time my husband was bitten at all was during the two days I was in Montreal attending a meeting of the Ecological Society of America; when I returned home he told me he had really missed me while I was gone.

After a week of this I was in terrible shape; between looking frantically around the house for evidence of fleas, scratching the unrelievable itching, and imagining symptoms of dog tapeworm infestation (humans can acquire the parasite *Dipylidium caninum* from fleas on their pets, although the usual mechanism of infection is by inadvertently swallowing the flea), I wasn't sleeping much more than two or three hours a night. Frankly, it's a mystery to me how cats, even when infested with fleas, manage to spend most of their time sleeping.

So, naturally, my judgment was impaired—at least, that's the excuse I offer for even briefly entertaining the idea of strapping flea collars around my ankles. I know it's a terrible idea, and I don't think I would actually have done it—the passing thought could have been a result of fleaborne murine typhus infecting my brain and altering my normally rational thought processes. But I was desperate—repellents weren't working at all. Neither DEET nor the newly available picaridin-based formulations provided any relief, and all that the herbal citronella-based repellents succeeded in doing was repelling my spouse, who claimed he couldn't sleep because of the smell.

Intellectually, I knew that flea collars are formulated for use on cats and dogs only. And I knew the products available in the late 1980s when I last encountered fleas were not ones I'd want next to my skin—collars for the most part contained either organophosphates (chlorpyrifos, dichlorvos, naled, tetrachlorvinphos, malathion, and diazinon among them) or carbamates (carbaryl), neither of which I was particularly anxious to place on my ankles, even if in doing so I could wreak havoc with the flea populations congregating there; their ability to interact with vertebrate nervous systems was disincentive enough. But I thought maybe, with the innovations in on-host therapy, a new generation of more user-friendly flea collars might be available. A quick search of the literature revealed that there's not much space devoted to flea collars. I did find a fascinating review of the history of collar technology (Witchey-Lakshmanan 1999) and learned that the first flea collar appeared in 1963 and consisted of liquid dichlorvos incorporated into a vinylic resin strip, marketed as Sergeant's Sentry. I discovered that Bayer holds patents for the use of polyurethanes and ethylene vinyl matrices. And that there are patents for chambered collars that allow fleas to enter and get mired onto an internal adhesive, after the fashion of a Roach

Motel. None of this information, though, was directly applicable to my problem. I also found a new medical condition to worry about (anisocoria or "flea collar pupil"—Apt 1995). This also wasn't very helpful.

So, like many other desperate individuals, I turned to the Internet. I was amazed by the diversity of flea collars available. In addition to the standard OP/carbamate collars of old, there are now nonchemical flea collars, containing various sorts of herbal products that are ostensibly repellent, along with ultrasonic flea repellents, none of which have any demonstrable efficacy. There are even glow-in-the-dark flea collars, glitter flea collars, and velvet flea collars, perhaps for formal pest-management situations. All of the collars with traditional pesticides included the standard disclaimer that "it is against federal law to use this product in a manner inconsistent with its labeling", but I was surprised that none specifically stated on the box, "It is a stupid idea to attach this product to your ankles." In this litigious era, there are warning labels on everything—there are hair dryers with labels that say, "Do not use while sleeping" and picture frames labeled, "Not to be used as a personal flotation device." The flea collars rather mildly informed me that they were "harmful to swallow" and that it would be a good idea to "avoid contact with skin," but I expected more specific warnings. And I began to wonder if I was the only one who had ever thought of the idea.

Well, of course I wasn't. That's about the time I read the warnings from the U.S. military and saw the graphic consequences of flea-collar abuse. Eventually, the flea problem abated enough that I recognized the magnitude of my momentary stupidity.

I'm speechless with admiration for my colleagues in veterinary entomology—the new products in use for flea control are nothing short of revolutionary. In fact, it would be nice if Advantage could be formulated for use on humans. Not that I even fleetingly considered applying a tube to my own shoulder blades (although, again, there's no warning on the box—just the usual mild direction to "avoid contact with skin"). I won't. And if you see me, don't even ask; "don't ask, don't tell" works for entomologists, too.

the jumping face bug

THE VAST MAJORITY of people in most regions are keen to keep their private parts private—that is, to keep the number of uninvited visitors to a minimum. Fetishism aside, finding any kind of arthropod in residence on any part of the body is likely to inspire an intense form of fear and loathing. This deep-seated distaste underlies a condition known as Ekbom's syndrome, or delusory parasitosis, a psychotic disorder characterized by the persistent, unshakeable, but false conviction that insects or other small crawling creatures are living in or on the skin. People suffering from this disorder have been known to scratch themselves until they bleed, inflict deep slices into their skin, and douse themselves with pesticides or gasoline in their fruitless efforts to rid themselves of even the thought of parasites.

In November 2004, the Entomo-l Listserv was abuzz with news about an unusual paper that had appeared in the *Journal of the New York Entomological Society* earlier in the year. The paper, titled "Collembola (springtails) (Arthropoda: Hexapoda: Entognatha) found in scrapings from individuals diagnosed with delusory parasitosis," was coauthored by six people, none of whom I knew personally but whose ranks included, among others, the commissioner of health from the Oklahoma State Department of Health, a faculty member from the Department of Parasitology in the University of Veterinary Medicine in Iasi, Romania, and a member of the scientific staff in the Division of Invertebrate Zoology at the American Museum of Natural History.

I probably should have anticipated that any entomological paper with the word "scrapings" in the title would be disquieting, but this paper (Altschuler et al. 2004) exceeded expectations in this regard. The authors reported the results of their microscopic analyses of skin scrapings from twenty subjects previously diagnosed with delusory parasitosis. Scrapings from eighteen of these twenty were reported to have flourishing populations of collembolans living in their skin, as evidenced by remnants of eggs, shed skins, and nymphs. Thus, according to the authors, the parasitosis was genuine and not delusory in 90 percent of the cases examined.

The fact that there are small, crawling animals that may escape detection and cause intense itching is always a lingering concern for entomologists confronted with individuals who claim they are besieged with small crawling animals. Some types of mites, for example, can come into a house with a squirrel or mouse and, after their host dies, seek out a human alternative. In the case of the Oklahoma study, some insect taxonomists might disagree on whether these individuals actually were hosting populations of *insects* in their skin, given the ambiguous status of the order Collembola—as wingless hexapods lacking the kind of respiratory apparatus possessed by all other insects, it's unclear whether the springtails belong in the same class. And the authors shed no light on whether these springtails were small *crawling* animals, or, as is their wont, small *hopping* animals (springtails owe their name to a forklike appendage that they slap against the ground to propel themselves several inches into the air). Notwithstanding, I was totally creeped out.

I'm one of those people who cannot read about a medical condition without immediately becoming convinced that I'm suffering from it. Over the years, I had barely become reconciled to the idea that follicle mites might be crawling around on my face; *Demodex follicularum* is a tiny mite that crawls into hair follicles, usually on the face, and feasts on sebaceous gland secretions. In truth, I've always held tenaciously to the belief that, because only three-quarters of the human population houses these mites, there's a chance I'm in the one-quarter that remains by virtue of clean living and dumb luck mite-free. The thought that there are hexapods hopping about seemed infinitely less tolerable.

Admittedly, I don't know a lot about collembolans—I don't regard them as true insects and hence they're not my responsibility—but what little I did know didn't jibe well with this report. Springtails, as the authors themselves describe, typically live in "moist environments and abundant organic debris," certainly not phrases used often in connection with human facial skin, unless this condition perhaps results from moisturizer abuse. The photos in the article were of little help; the images were difficult to discern, and I haven't looked at too many collembolans under the microscope. There were arrows pointing to vague, dark shapes labeled "furcula" and "collophore," which I knew to be collembolan body parts, but I was sufficiently unsettled by the thought that I might someday have springtails jumping on my face that I couldn't bring myself to look more closely.

Apparently, other entomologists, far more secure with the status of their facial fauna, were not unwilling to scrutinize the images. This article hit the Entomo-l Listserv in November, 2004, and generated a heated discussion over whether the images were real or the result of well-meaning or misguided computer enhancement—the scientific equivalent of the image of the Virgin Mary appearing in a grilled-cheese sandwich that was miraculously preserved for a decade before being auctioned off on eBay for $28,000 to an online casino. In other words, collembolans in skin scrapings might be just another example of the phenomenon of pareidolia, "a type of illusion or misperception involving a vague or obscure stimulus being perceived as something clear and distinct" (Carroll 2005).

Pareidolia is a widespread phenomenon and is in fact part of human culture. There's a distinct, nearly universal human predisposition to see human faces everywhere—the "man in the moon" being a case in point. Carl Sagan even went so far as to suggest that such a predisposition might even be adaptive: "As soon as the infant can see, it recognizes faces, and we now know that this skill is hardwired in our brains. Those infants who a million years ago were unable to recognize a face smiled back less, were less likely to win the hearts of their parents, and less likely to prosper. These days, nearly every infant is quick to identify a human face, and to respond with a goony grin" (Sagan 1995: 45).

Entomologists are not immune from the phenomenon; seeing bugs on human faces is thus an odd mirror image of the more typical pattern of seeing human faces on bugs. There is a remarkable array of species named for their imagined resemblances to human faces of various descriptions. Among the friendlier manifestations of this phenomenon is the Hawaiian happyface spider, *Theridion grallator,* a comb-footed spider found in Hawaii named for the smiley-face markings on its abdomen. Other arthropod faces are far less jovial. There are dozens of species with common names that make reference to more foreboding faces—the death's head cockroach, for example, or the skull and crossbones roach *Blaberus craniifer* (although, given that the skull and crossbones pattern is also known in the pirate world as the "Jolly Roger," I suppose it could be considered a happy face of sorts).

Undoubtedly, the most famous of the insects sporting human faces are the death's head hawk moths in the genus *Acherontia*. These moths all have an eerily realistic skull pattern on the thorax. The universality of the tendency to see human faces (albeit skeletal ones) is evidenced by the common names of *Acherontia atropos* throughout Europe, which translate literally to "death's head" in Czech, Danish, Dutch, Estonian, Finnish, French, German, Hungarian, Polish, Russian, Spanish, and Swedish.

The adaptive value of a skull-and-crossbones marking has long been a puzzle; the Jolly Roger and the universal symbol for poison arose long after hawk moths evolved. Miriam Rothschild (1985) suggested that the markings on the thorax of the death's head moth depict not a human face but rather the face of a honey bee queen, allowing the moths unmolested entry into the hives of honey bees, where they use their stout proboscis to pierce sealed comb and steal honey: "in the obscurity of the hive, and seen from above by the guard bees, . . . the "face" set immediately above the brown and yellow striped segmented abdomen, topped by the bee's own antennae could give the impression of a huge gravid queen bee; a super model, if ever there was one."

Although Kitching's (2003) take on the death's head moth is that there is "no entomologist, anthropomorphic or otherwise, who actually believes this pattern was meant to represent a human skull," the argument has been made in at least one case that markings resembling the human face are in fact the product of natural selection rather than pareidolia. H. E. Hinton (1974) made the case that pupae of several lycaenid butterfly species in Africa gain protection from their resemblance to "the head of a monkey" because "to some birds the pupa sufficiently resembles a monkey so that at least in a small percentage of instances avoiding action is taken with the result that the pupa escapes attack." The resemblance of the New World harvester butterfly *Fenisca* [sic] *tarquinus* to an Old World anthropoid ape is more difficult to explain in the absence of any anthropoids in the New World other than humans, who "meet a further requirement of the presumed model, namely, that of being harmful to birds." Faces may even deter human predators; Hinton (1973) cites Huxley's 1957 speculation that the Japanese crab *Dorippe japonica* resembles "an angry traditional Japanese warrior" as the result of natural selection—"the resemblance . . . is far too specific and far too detailed to be merely accidental; it is a specific adaptation which can only have been brought about by means of natural selection operating over centuries of time, the crabs with more perfect resemblances have been less eaten." Apparently, vegetarians aren't the only ones who won't eat anything with a face.

Japan seems to be more fertile ground than most places for pareidolia. Chonosuke Okamura, a twentieth-century Japanese paleontologist, was perhaps the poster child for the phenomenon. Where his colleagues saw fossil fragments of coral and microbes in limestone deposits, Okamura (1980, 1983) saw over ninety species of tiny vertebrates, all of whom were identical in every detail to contemporary vertebrates except for size. He described in excruciating detail an entire civilization of Silurian-era "protominimen" identical to contemporary humans except in stature, declaring "There have been no changes in the bodies of mankind since the Silurian period . . . except for a growth in stature from 3.5 mm to 1700 mm." Okamura described an entire civilization of *Homo sapiens miniorientalis,* accompanied by their domesticated mini-dogs *(Canis familiaris minilorientalis),* worshipping mini-dragons *(Fightingdracuncus miniorientalis),* and otherwise engaging in the same range of cultural activities as their gargantuan descendants (including dancing—one photo purports to show "two totally naked homos, facing each other . . . moving their hands and feet harmoniously on good terms. We can think of no other scene than dancing in the present-day style"). Given that the minimen were "about the size of an aspirin tablet" (Abrahams 2002), the level of detail in what Okamura imagined he saw was remarkable. All the same, I suppose it's understandable that he failed to discern or omitted to mention, in his voluminous publications on the subject of minimen, whether they had to deal with minispringtails on their faces.

the kissing bug

EVEN THOUGH MOST people are willing to believe the worst about insects, there are some insects that just seem to stretch the imagination beyond credulity. Kissing bugs are a good example. For more than a century people have doubted the existence of a bug that, under cover of night, eschews every other part of the human body and heads directly to the lips to inflict a painful bite. On a Web site called Cracked.com, there is a long thread about people being killed by gravy boats. It contains this contribution, presumably regarded as equally improbable:

"The South American Kissing Bug will climb onto your face while you are sleeping, eat your lips and [defecate] on your face."

Although juxtaposition with reports of death by tableware would tend to promote skepticism, it's actually quite true that any of a number of South American kissing bugs and, for that matter, kissing bugs from other continents will indeed climb on your face while you're sleeping, eat (or at least bite) your lips, and defecate on your face. The term "kissing bug" is a common name for a number of species of bloodsucking parasites in the generally predaceous bug family Reduviidae. The predilection of certain species for sucking blood from the faces of sleeping humans gave rise to the common, albeit somewhat inappropriate (depending on personal proclivities for expressing affection), name. The predaceous members of the family, which sink their proboscis into just about any part of the body of fellow arthropods to suck their blood, are known by the far less romantic common name "assassin bug." Kissing bugs don't have particularly stout or strong mouthparts, so they take the easy route to a blood meal and pierce the thinnest skin they can find—the skin of baby rodents in underground burrows. If they stumble across a sleeping human above ground, they'll find blood through the thinnest skin available, surrounding the eye or on the lips. Due to a combination of stealth and anesthetizing saliva, their feeding rarely wakes their sleeping victim, who awakens with otherwise inexplicably puffy lips or eyelids.

A particularly obnoxious South American representative, *Triatoma infestans,* is locally known as the vincucha. Its kiss can be the kiss of death, because this insect is the principal vector, or carrier, of a one-celled protozoan pathogen known as *Trypanosoma cruzi,* an infectious microbe that is the cause of the debilitating illness called Chagas disease, or American trypanosomiasis.

Infection results not from an injection of the microbe through the proboscis (as is the case for malaria, yellow fever, and other arthropod-borne scourges); infection is in a way self-inflicted in that the pathogens are contained in the droppings of the kissing bug, which fall onto the skin and get rubbed into the feeding site when a victim scratches an itch.

Chagas disease is particularly insidious in that symptoms of infection may not manifest themselves for decades; up to 30 percent of infected individuals can develop chronic symptoms culminating in heart failure. Short of heart failure, Chagas disease can also cause a whole spectrum of symptoms, including but not limited to persistent fevers, fatigue, anemia, swollen lymph nodes, and a characteristic one-sided facial swelling called a chagoma (or sign of Romaña). Over 11 million people suffer from Chagas disease and over 20,000 of these die every year (Dias 1997; Moncayo 2003).

So there's good reason to fear *T. infestans* and its fellow vectors (which number over a dozen in the genus *Triatoma* alone). Many other species can suck human blood with no more serious consequences than a temporary fat lip. For some reason, though, the implicit intimacy of mouth-to-mouth contact is unnerving to many. In the summer of 1899, a kind of kissing bug mania seized the nation. It seemed to start in Washington, DC and moved quickly to Brooklyn and then to New York, where the *New York Times* ran almost daily stories about new victims of kissing bug attacks. On July 3, a kissing bug "descended upon Atlantic City and touched the lips of Mrs. Helen Veasy, a cottager at 213 Chalfront Avenue, and John McCaffrey, a fourteen-year-old Western Union messenger boy. Mrs. Veasy was sitting last night on the porch of the cottage when a large insect came sailing up to her in the semi-darkness and lighted for an instant on her lip. As she raised her hand to brush it away she felt a slight pain shoot through her upper lip." Within a half-hour her lip had swollen to the size of a "robin's egg." The messenger boy woke up with a lip so swollen he "was scarcely able to masticate his food." The story concluded with a report that a kissing bug caught the day before by a trolley car motorman was "displayed today in a store window" for all to see. On July 7, a report came in from New Rochelle of two more victims, a toddler and a teenager; on July 11, four patients were treated at Bellevue Hospital in Manhattan, with lips variously swollen to two, three-and-a-half, and four times normal size.

By July 14, doubt had begun to set in; in the "Topics of the Times" column there appeared an editorial:

> this and other papers are recording daily the suffering of people who think that they have been bitten by a newly invented insect. That the bites are inflicted by something is probably true, and there is no doubt at all about the tumefaction and the pain which the doctors are called upon to reduce and assuage and yet there is not the faintest evidence to show . . . that a strange insect has made its appearance in the country. . . The entomologists unite in denying the existence of a "kissing bug," and though little creatures by the dozen of one sort or another have been submitted to these authorities as criminals caught in the very act of osculatory attack, still the learned ones refuse to be convinced, and not only give the accused names as ancient as they are long, but assert and, if need be prove that the incriminated bugs couldn't bite if they would and wouldn't if they could. The whole trouble seems to be the effect of misguided and overexerted imaginations . . . This "kissing bug" epidemic tends to show that "faith" can cause functional disturbances as well as remove them. But everybody knew that before.

After several more days of attack reports, finally, on July 20, no less an authority than the chief entomologist of the U.S. Department of Agriculture, Leland O. Howard, weighed in on the kissing bug epidemic. In a story titled, "The *Opsecoetes personatui*: That is the Kissing Bug's family name and Dr. Howard says his bite of itself is not dangerous," the *New York Times* reported that

> a new terror is added to the kissing-bug craze by Dr. Howard, who declares that we have got to unlearn the name "pici pes" and practice on that of *Opsecoetes parsonatui*. It seems that "pici pes" is simply an alias under which the bug has been masquerading.

A second story ran on August 19 with the headline, "Kissing Bug Is Not a Myth." Howard's point was that the bites were likely inflicted by *Reduvius personatus,* the masked bed bug hunter, an assassin bug that normally attacks bed bugs and thus typically imbibes human blood only from the stomachs of bed bugs, but for reasons unknown to entomology was uncharacteristically abundant at the moment and by sheer force of number was biting humans in its nightly search for bed bug blood.

Although one would think that this information would have settled the matter, reports of kissing bug attacks continued to appear for the rest of the summer from across the country (including Rhode Island by July 22 and Peoria, Illinois by July 30). The reports grew progressively stranger with time; a story from July 28, for example, reported on an attack on Frank, a performing polar bear on Coney Island, and a story from September 10 of a British freighter returning from Bombay reported on "two monkeys that had come aboard in India [who] withstood the attacks of the kissing bugs for a week and then decided that a peaceful death by drowning was preferable to going through life with swollen lips . . . So both monkeys leaped overboard from the steamer two days before the entrance of the Suez Canal was reached."

It wasn't long before kissing bugs began to appear in the entomological literature. First, and perhaps prematurely, was a report in the 1899 "Annual Report of the Entomological Society of Ontario"; according to H. H. Lyman, the president of the society,

> One other event of the past season to which I should refer, was the advent through the medium of the daily press, of a terrible bogey in the form of a bloodthirsty insect which was "written up" by the knights of the quill under the name of the Kissing Bug. It was said that its scientific name was Melanolestes Picipes, and the wildest stories were told of its deadly ravages. Illustrations of it were published, and various kinds of insects of different orders were exhibited in newspapers' windows as genuine specimens of the bug. Quite a number of deaths were attributed to it, and many timid people, especially women, were seriously alarmed. It started from Washington (there is something very suspicious about this, but perhaps our friends of the Division of Entomology can establish an alibi) and spread all over the continent, creating devastation everywhere with the exception, it is said, of Baltimore, whose newspaper men are reported to have been too conscientious to write it up, though the latter statement seems almost more incredible than the stories told of the bug. At last the secret was given away and the kissing bug pronounced a myth, the story having been started as a hot weather silly season hoax. (Lyman 1899)

Lyman then thanks one "Dr. Howard for his kindness in favoring me with much interesting information and valuable suggestions." This is despite the fact that Dr. Howard clearly came out in support of the reality of kissing bugs months earlier. By November 1899, Howard had compiled a list of six reduviid species most frequently identified as "kissing bugs" the preceding summer, publishing it in *Popular Science Monthly* and re-issuing it a decade later as a USDA bulletin (Howard 1899).

Once the night terror had a name, the mania subsided, but the kissing bug had a lasting impact on popular culture. The anthology *Love Among the Mistletoe, and Poems,* published in 1908 by James Buchanan Elmore, included a poem titled "Kissing Bug." Although the central thesis is that some "kissing bugs" are men who smell "like foaming beer," some entomological information appears in a verse or two:

> Some ladies are afraid of a kissing bug
> And cannot sleep o' night
> And yet they embrace and kiss a thug
> And think it out of sight.
> This bug appears when snug in bed,
> And you are sound asleep;
> You'll feel it crawling o'er your head,
> And touch your rosy cheeks . . .
> You'll know this bug, with tweezers sharp,
> And beak that's very black;
> You'll feel so queer about the heart
> As he takes a dainty smack. (p. 114)

In 1909, the kissing bug was immortalized in song by Charles Johnson, the ragtime composer, in the "Kissing Bug Rag." The sheet music unfortunately didn't include any images of kissing bugs, but instead pictured a winged woman surrounded by tiny winged male suitors in tuxedos). Shifting comfortably with fads in musical genres, kissing bugs were also featured in the "Kissing Bug Boogie," written by Charles "Crown Prince" Waterford in 1950.

It's likely that songwriters who immortalized the kissing bugs were inspired only by the name, but they might have been interested to know that some kissing bugs can actually sing themselves (Schofield 1977). A number of reduviid bugs, including bloodsuckers called cone-nose bugs, are capable of stridulating, or making a kind of chirping sound, by rubbing the rigid tip of their mouthparts against a series of ridges on the underside of the thorax. Why they do this is subject to some discussion—the consensus is that stridulation is a defense against predators. At least one competing hypothesis, however, led to yet another appearance of kissing bugs in popular culture. According to a front-page story in the U.K. *Guardian* that ran on June 7, 1966, the U.S. Pentagon was reported to be

> planning to send bed-bugs [sic] to help to win the war in Vietnam . . . Their plans are based on the fact that bed-bugs scream with excitement at the prospect of feasting on human flesh . . . a sound amplification system would enable the GI, sweating through the jungles of South Vietnam, to hear the anticipatory squeals of a captive bed-bug as it detects the Vietcong lying in ambush ahead. Tests have apparently shown that a large and hungry bed-bug will appropriately register the presence of a man some two hundred yards to its front or side, while ignoring the person carrying it in a specially devised capsule.

There are so many biological improbabilities in this account that it's hard to believe that such a proposal was ever seriously considered by the Department of Defense—although it must be noted that this same government agency was accused by the Nazi government in 1942 of plotting to drop 15,000 Colorado potato beetles onto German potato fields to destabilize German food supplies (leading to the establishment of a Kartoffelkafer-abwehrdienst, "Potato Beetle Defense Service"). Not to be outdone, the government of the German Democratic Republic in 1950 accused the United States of actually dropping thousands of Colorado potato beetles in the southwest part of the country to demoralize a nation fond of its potato dumplings (Garrett 1996). Most estimates of the distance from which blood-sucking bugs can detect a human blood meal are in the range of ten to twenty feet, not 200 yards. Moreover, it's unclear how or why bugs would ignore a potential blood meal almost underfoot in preference to a meal two football fields away. There's no evidence that bugs can differentiate between allies and hostile forces, either. Finally, what adaptive value there might be to a bug of announcing its presence while stalking its prey by squealing in anticipation is a mystery; it would seem such squeals are completely inconsistent with the stealth strategy displayed by the group as a whole. If nothing else, any Viet Cong within earshot would know of the presence of an American soldier immediately upon hearing the squeal. It would seem, then, that the amplified scream of a kissing bug could instead be a signal for the American soldier to kiss his chances goodbye.

the mate-eating mantis

EVEN PEOPLE WHO may be uncertain as to how many legs an organism can have and still be considered an insect know one insect fact with certainty—that, in the act of mating, the female praying mantis kills and eats her partner, usually devouring the head first. This bit of biology has been celebrated in virtually all forms of written expression. It shows up a lot in screenplays, to which it is ideally suited. How many other metaphors evoke sex, murder, decapitation, and cannibalism? That female mantids eat their mates is likely the most widely known entomological fact. Unfortunately, this fact probably isn't true, at least in the vast majority of cases.

Why isn't it generally true? For one thing, there are over 2,000 species of mantids in the world and the phenomenon has been reported in only a tiny handful of them. Secondly, most reported cases have involved captive specimens and the sexual cannibalism was likely a laboratory artifact. Mantids kept in captivity are highly constrained by the circumstances of their confinement and are more likely to be chronically hungry or malnourished, given the rather primitive state of knowledge of mantid minimum daily nutrient requirements. Another factor contributing to the notion that sexual cannibalism is commonplace is that it's a natural human tendency to remark upon and remember extraordinary sights; seeing mantids in copulo that parted ways amicably after the deed was done is hardly worth noticing.

This particular myth likely became pervasive because it happened to be remarked upon by a handful of extremely eloquent writers in high-profile places. The first reference in the peer-reviewed scientific literature dates back to 1886; it didn't hurt that it was published in *Science,* the premier scientific journal of the era, and that it was written by Leland Ossian Howard, future chief entomologist of the U.S. Department of Agriculture.

> I brought a male of *Mantis carolina* to a friend who had been keeping a solitary female as a pet. Placing them in the same jar, the male, in alarm, endeavored to escape. In a few minutes, the female succeeded in grasping him. She first bit off his left tarsus, and consumed the tibia and femur. Next she gnawed out his left eye. At this the male seemed to realize his proximity to one of the opposite sex, and began to make vain endeavors to mate. The female next ate up his right front leg,

and then entirely decapitated him, devouring his head and gnawing into his thorax. Not until she had eaten all of his thorax except 3 millimeters did she stop to rest. All this while the male had continued his vain attempts to obtain entrance at the valvules, and he now succeeded, as she voluntarily spread the parts open, and union took place. (Howard 1886)

The story was rendered even more eloquently eleven years later by the great popularizer of insects, Jean Henri Fabre: "if the poor fellow is loved by his lady as the vivifier of her ovaries, he is also loved as a piece of highly flavored game . . . I have . . . seen one and the same mantis use up seven males. She takes them all to her bosom and makes them pay for the nuptial ecstasy with their lives" (Fabre 1916). Whatever the story lacked in scientific vigor it more than made up for in flowery and evocative language. Despite having worked with insects for three decades, I can't begin to tell you exactly where one might find a "bosom" on a mantid. Finally, in 1935, Kenneth Roeder, a physiologist, wrote a paper that catapulted the factoid into iconic status. Roeder knew that male mantids routinely mate and escape to mate another day and even remarked upon this in his paper. What attracted the most attention in the paper was his observation of one particular copulation that went awry. As this male was being decapitated, his twisted genitalia flipped around and engaged the female; ultimately the act was completed without benefit of forethought or afterthought (i.e., without benefit of a brain). This observation led Roeder to suggest that the subesophageal ganglion, the principal nerve bundle in the head, might send inhibitory signals to the abdomen. Removal of the head thus may free up inhibitions. He also reported a year later that removal of the head of the female also activates her genitalia, but this particular observation, lacking the pat adaptive explanation, never really had much impact.

There are many other explanations as to why head removal leads to genitalic torsion in both sexes of mantids. It's widely known that various sorts of damage to the nervous system of vertebrates can lead to aberrant motor behaviors because of the injury-related release of inhibitory signals (Kandel and Schwartz 1985). A spectacular manifestation of this phenomenon are the reflexive erections that can be induced in human patients subjected to spinal block or some kinds of brain lesions. Despite the fact that this response is well known, nobody is suggesting that it is in any way adaptive nor is it likely that brain lesions will be recommended any time soon in marital aid manuals.

The truth of the matter is, if you're a small arthropod with the noblest of intentions, walking up to a larger carnivorous arthropod, even a member of the same species, is a tricky business. Many carnivores orient specifically to motion, and a prospective mate with all of the wrong moves around a hungry female puts not just his head but virtually every other edible body part at risk. That this is the norm is suggested by the fact that males of many carnivorous species, particularly those in which males are significantly smaller than females, have all kinds of ways of reducing the probability that their intentions will be misinterpreted. The European mantis *Mantis religiosa*, for example, moves six times faster toward a female whom he observes eating or holding prey. Seeing her eat her fill is evidently a risk-reducing turn-on (Gemeno and Claramunt 2006). Males of the Chinese mantis *Tenodera aridifolia sinensis* are significantly less likely to court hungry females; when they do court them, they court with more exaggerated movements (described by one pair of authors as "upward thrusting of the forewing and hindwings and a rhythmic bending motion of the abdomen") and mount them by leaping onto their backs from a greater distance (Lelito and Brown 2006). The male fishing spider *Pisaura mirabilis* tries to distract his carnivorous mate with a dead fly or similar prey item, often wielding it as a shield as he approaches; in the often overheated vocabulary of ethology, these are called "nuptial gifts," although they're hardly likely to show up on any registry at Macy's. If the female attacks him instead of the gift, he plays dead until she shifts her attention back to the prey item (Bilde et al. 2006).

Among arthropods, it really is an eat-or-be-eaten world; examples of all kinds of arthropod cannibalism, including sexual cannibalism, abound, but they are rarely objects of conversation. Without the element of decapitation, they fail to capture the public's imagination. Horned nudibranchs, colorful albeit amorphous sea slugs, display sexual cannibalism, but it's difficult to tell upon casual inspection exactly where the head is. And there are male insects that offer up various and sundry nonhead body parts or secretions to make females less hungry and more inclined to mate. The male speckled cockroach *Nauphoeta cinerea* produces a pheromone called "seducin" in an array of glands on his back and sides that the female licks while he plans his approach (Sreng 1990). But some females don't stop at just a lick—they actually eat bits and pieces of the males they mate with. Female sagebrush crickets in the genus *Cyphoderris* eat the fleshy hind wings of their mates and assiduously lick the blood (or, technically speaking, hemolymph) that flows from the wounds they inflict (Johnson et al. 1999). In the anthicid beetle *Notoxus monoceros,* it's the male's anal sacs that the females nibble on; females can consume up to several percent of their body weight in their prospective mate's more private parts. Ostensibly, these anal sacs provide not only nutrients but also valuable chemicals that females can redirect after ingesting them into their eggs to protect them against predators. It's really quite a reasonable arrangement, at least for the beetles. Somehow, though, even if it cuts down on the expenses, it's hard to imagine that, in human courtship, anal sacs will replace engagement rings any time soon.

the "locust"

EUROPEANS HAVE LONG been convinced of the superiority of virtually every dimension of their culture relative to the American equivalent. Surprisingly, though, their initial impression of even American natural resources was, at least in some circles, disdainful. Georges-Louis Leclerc, comte de Buffon, a French naturalist of note in the eighteenth century, was convinced in fact that the New World was a wasteland and all of its organisms were smaller, less diverse, and in every way inferior to their European equivalents. In his *Histoire naturelle, générale et*

particulière, Buffon presented his "Theory of American Degeneracy":

> In America . . . animated Nature is weaker, less active, and
> more circumscribed in the variety of her productions; for we
> perceive, from the enumeration of the American animals,
> that the numbers of species is not only fewer, but that, in gen-
> eral, all the animals are much smaller than those of the Old
> Continent.
>
> No American animal can be compared with the elephant,
> the rhinoceros, the hippopotamus, the dromedary, the cam-
> elopard [giraffe], the buffalo, the lion, the tiger, &c . . . Hence
> in the New Continent, there are more running waters, in pro-
> portion to the extent of territory, than in the Old; and this
> quantity of water is greatly increased for want of proper
> drains or outlets . . . Besides, as the earth is everywhere cov-
> ered with trees, shrubs, and gross herbage, it never dries. The
> transpiration of so many vegetables, pressed close together,
> produce[s] immense quantities of moist and noxious exhala-
> tions. In these melancholy regions, Nature remains concealed
> under her old garments, and never exhibits herself in fresh
> attire.

All that muck and mire, however, was apparently great at generating lower life forms—"insects, reptiles, and all the animals
which wallow in the mire, . . . and which multiply in corruption,
are larger and more numerous in the low, moist, and marshy
lands of the New Continent." In short, noble beasts were no-
where to be found but anything vaguely creepy, multi-legged and
slimy felt right at home in the degenerate New World. In contrast with European scholars, New World settlers may have been
more impressed. The appearance of millions of strange, noisy

six-legged creatures in Plymouth colony in 1634 led the colonists to believe they were experiencing a plague of biblical proportions. Later colonists, on hearing the accounts, called the creatures "locusts," even though they bore little to no resemblance to the migratory grasshoppers of Biblical fame. Even the hard-to-impress Europeans didn't know what to make of this American phenomenon—one of the earliest descriptions appeared in 1666, in the first volume of *Philosophical Transactions of London,* in a paper titled "Some observations of swarms of strange insects and the mischiefs done by them." As it turns out, Europeans were nonplussed by these American insects because they have nothing equivalent to American periodical cicadas.

Although there are about 1,500 species of cicadas in the world, all largish insects with clear wings that feed with a beaklike proboscis on the dilute sap of plant roots and make buzzing sounds by vibrating drumlike organs called tymbals on their abdomens, only seven American species, the periodical cicadas, can lay claim to having the longest juvenile developmental period of any insect. Depending on the species, the transition from egg to adult takes either thirteen or seventeen years. Moreover, because these thirteen- or seventeen-year cycles run synchronously in different geographic locations, within ten to fourteen days of each other literally millions of individuals of a given population, or brood, tunnel up out of the ground, shed their last nymphal skin, and celebrate adulthood by singing to attract a mate, in the case of the males, and by laying from 400 to 600 eggs, in the case of the female. The simultaneous song stylings of millions of males can reach deafening proportions—in some places, eighty to ninety decibels in intensity, equivalent to the noise level of a subway station.

Nineteenth-century American entomologists carefully mapped the distribution and abundance of the various and sundry populations of these insects, which run on different thirteen- or seventeen-year cycles, depending on species. One in particular, Charles Marlatt, hit upon the idea in 1898 of giving the different populations Roman numerals. Numbers I through XVII were assigned to seventeen-year cicadas and XVIII to XXX to thirteen-year cicadas (although subsequent entomologists determined he had overestimated brood numbers).

Despite the predictability of these emergences, people always seem to be caught by surprise when particularly large broods emerge. Brood X, for example, comprises much of the eastern half of the United States. When this brood emerged in 2004, television reporters and radio personalities, not knowing or remembering the entomological convention, repeatedly referred to it as Brood X (as in, "letter that precedes Y").

With this massive spring emergence of Brood X periodical cicadas in Washington, DC, I guess even the most insect-averse Washington politico couldn't fail to notice them. By sheer force of numbers, Brood X made an impact on the cultural scene, drowning out weddings, clogging pool filters, appearing on t-shirts and hats, showing up in stir-fries and in smoothies by design as well as by accident, and otherwise making their presence known, so it was probably inevitable that they'd blunder into partisan politics. First to take metaphorical advantage of the infestation was the Republican National Committee. On May 14, 2004, at the height of the emergence, 700,000 registered Republicans received an email attachment from the Republican National Committee. A narrator intoned, "Every 17 years, cicadas emerge, morph out of their shell, and change their appearance. The shells they leave behind are the only evidence they were here. Like a cicada, Senator Kerry would like to shed his Senate career and morph into a fiscal conservative, a centrist Democrat opposed to taxes, strong on defense . . . But, he leaves his record behind . . . when the cicadas emerge, they make a lot of noise. But they always revert to form, before disappearing again." The voiceover accompanies a time-lapse film of a cicada eclosing and expanding its wings and ends with an animated cicada morphing into John Kerry.

The Kerry campaign wasn't overly bothered by the advertisement; in fact, a representative told a *Cincinnati Enquirer* reporter that the campaign wasn't "bugging out" over the advertisement (adhering to the time-honored journalistic tradition of accompanying news stories about insects with laborious puns) and added that, "Maybe, if given another 17 years, President Bush could create a job in Ohio" (Korte 2004). As an entomologist, I confess to being slightly baffled by the use of insect imagery to promote a political candidate. If I'm for Kerry, does that mean I'm against cicadas? Are cicadas Republican or Democrat? Do other insects have party affiliations? As far as metaphorical metamorphic transformations go, I don't exactly get it, either—cicada nymph to cicada adult is hardly the most dramatic. Were I picking campaign metaphors, I might have gone with grub to beetle, or maggot to fly; there's lots more metaphorical power there.

But then, maybe I just don't know enough about cicada-related political history. Even though the Brood X emergence in the Baltimore-Washington area coincides with a presidential election only every sixty-eight years, since the nation's capital was moved there July 16, 1790, area Brood X cicadas have been destined to cross paths with politicians whenever they emerged. Gene Kritsky, noted cicada authority, pointed out to a press corps interested in insects about once every seventeen years that cicadas on at least one other occasion had had an impact on presidential politics. Back in 1902, President Theodore Roosevelt was practically drowned out while trying to give a Memorial Day speech defending national policy to impose "orderly freedom" in the Philippines.

This is not to say that cicadas are the only six-legged strange bedfellows of politicians. Insects played a critical role in a bitterly contested presidential election over a hundred years ago. In 1896, Republican William McKinley faced Democrat William Jennings Bryan and embroiled the nation in a dispute over U.S. monetary policy—specifically, whether gold or silver should serve as the national standard. The Republicans adamantly opposed the free coinage of silver and maintained that all coins and paper maintain parity with gold. For their part, Bryan and the Democrats were unalterably opposed to this policy and argued passionately for the free and unlimited coinage of silver and gold. It happened that, three years earlier, an outfit called Whitehead and Hoag filed a patent for what is now called a campaign button (technically not a button at all but rather more of a pin, and therefore called a pinback button). For reasons lost to history, both parties decided to display their loyalty to their candidates with campaign buttons shaped like insects. These pins, called variously gold bugs or silver bugs, depending on party affiliation, weren't true bugs at all—some looked like bees and others like stag beetles with misshapen mandibles. Bryan's silver bugs were often bedecked with such slogans as "Bryan for the bug house" and "How the farmer loves gold bugs." Pins from both parties often had photographs of the candidates and their running mates on the wings of the bugs, which folded up and popped out when the stinger was pushed. (Adlai Stevenson, the grandfather of the two-time unsuccessful presidential candidate with the same name in the 1950s, was Bryan's running mate and Theodore Roosevelt was McKinley's.)

Even third-party candidates got in on the insect action in the 1896 election. Democrats who disagreed with the party platform and embraced the gold standard broke away to become the "Gold Bugs" or "Gold Democrats"—they even went so far as to have their own nominating convention and put forward John M. Palmer, a 79-year-old Kentuckian as their candidate. Palmer was ignominiously defeated (accompanied, no doubt, by a wash of ponderous insect-related puns in the press).

The 1896 election apparently began a longstanding tradition of denoting unusual political associations with insects. Moderate Republicans in Congress from northeastern or midwestern urban states, for example, have been known as "gypsy moths" ever since a handful supported the impeachment of Richard Nixon in 1973. In contrast, conservative white Democrats from southern states with agricultural constituencies have been called "boll weevils" since the early 1980s, when Representative Marvin Leath from the eleventh district in Texas founded a conservative Democratic faction that allied with Republicans on tax and spending bills. As evidence that politics has indeed gotten uglier, it's worth noting that in neither case was the insect appellation chosen with affection.

It's telling that insect-related political metaphors almost always seem to involve odd alliances or strange bedfellows. Maybe the general feeling is that the apparent incongruity of partnering insects with politics symbolically conveys odd alliances. But maybe people should look a little deeper to find the underlying natural connections. As Gore Vidal once pointedly noted, "Politics is made up of two words, 'poli,' which is Greek for 'many,' and 'tics,' which are bloodsucking insects." Although the entomology leaves a lot to be desired, the etymology certainly has its merits.

the nuclear cockroach

IT'S ONE OF THOSE redoubtable entomological truisms—if there's a nuclear war, cockroaches will be the only survivors. Cultural references to the radiation resistance and durability of cockroaches abound. According to an interview with Coby Dick, a member of the metal/punk group Papa Roach, the band picked the name because "a cockroach can survive anything: earthquake, nuclear holocaust. They come in small numbers, and then they infest. We want to infest the world." When officials in New Zealand wanted to alert the public about Y2K preparedness, they chose as their symbol a cockroach, because their ability to survive any kind of disaster, including a nuclear one, is so well known. And there's even a Web site advertising a "giant deluxe cockroach," a five-inch rubber roach "which like most cockroaches could probably survive a nuclear war." People can take comfort in the thought that the fortunate few who manage to escape nuclear Armageddon can play postapocalyptic practical jokes on one another.

It's not easy to figure out where this idea of cockroach radiation resilience originated. It doesn't seem to have been the result of an empirical test. Atomic bomb blasts have been (mercifully) few and far between. I couldn't find any obvious references to postblast investigators noting the presence of unperturbed roaches at Ground Zero. There are, however, dozens of references on the Internet and elsewhere to the handful of gingko trees that remained standing after the bomb blast in Hiroshima; vendors of herbal medicines unfailingly mention that fact when touting the virtues of gingko in maintaining mental acuity, preventing asthma attacks, speeding poststroke recovery, and increasing blood flow to the penis.

So, most likely, the idea that cockroaches would be the sole survivors of a nuclear holocaust must have come from laboratory studies on radiation resistance. But the existing laboratory data aren't exactly consistent with cockroach supremacy. Among other things, cockroaches are relative newcomers to the ranks of the radiation resistant. Probably the first study to determine the effects of radiation on insects dates back to 1919, when W. P. Davey tested the effects of small doses of X-rays on the longevity of the flour beetle *Tribolium confusum;* surprisingly, Davey found that chronic exposure to X-rays at a dose of about 60 rad actually prolonged the life of the flour beetles. This finding evidently languished in the literature for about thirty-seven years, until a dubious J. M. Cork (1957) repeated the study under more controlled conditions and, to his dismay, obtained essentially the same result. He concluded his paper with the remark, "It is hoped that the results reported on a simple structure of this kind will not be construed as a license for X-ray practitioners to become less critical of recognized safety factors in dealing with the human organism."

Perhaps the most widely cited study documenting effects of radiation on insects was H. J. Muller's (1927) demonstration of artificial transmutation of genes in *Drosophila melanogaster;* Muller was essentially the first person to induce mutations, an accomplishment that revolutionized the science of genetics. Hanson and Heys (1928) heralded the accomplishment as "one of the most notable events in the field of pure biology in this century" and extolled the virtues of the new mutagenic agent; "on one Sunday afternoon forty mutations were found. Prior to the use of the X-ray, if one mutation were found in forty Sunday afternoons the time would have been considered well spent." Exposure to X-rays, radium, and other sources of radiation thus replaced the multifarious harmful agents that had been tested and found wanting as mutagens, including, among other things, continuous and intermittent rotation, very high temperatures, very low temperatures, and in one failed attempt to mutagenize white rats, ten successive generations of daily exposure to alcohol fumes ("the young of ten generations of alcoholic ancestors were both physically and mentally the equals of the controls and in some cases slightly superior"). Muller's finding inspired a generation of biologists to expose all kinds of insects to radiation in order to induce mutation, including parasitic wasps. (Many of these studies, by the way, were funded by a grant from the Committee for Investigation of Problems of Sex of the National Research Council; the fact that this committee no longer exists at the NRC suggests that this august body eventually did work out its problems of sex, whatever they may have been).

It really wasn't until the 1950s, when peacetime uses for atomic radiation (particularly for by-products of the nuclear power industry) were at a premium, that radiation resistance in insects became a focus for research. In 1957, two reports appeared in the same issue of *Nature* that documented the use of gamma radiation for control of wood-boring insects as well as for stored-product pests; thus, among the irradiated wood borers were the common furniture beetle, the deathwatch beetle, and the powderpost beetle, and among the irradiated pantry pests were the rice weevil, the grain weevil, the lesser grain weevil, the red flour beetle, the confused flour beetle, the sawtoothed grain beetle, the Khapra beetle, the cowpea weevil, the tobacco moth, the Mediterranean flour moth, and the Angoumois grain moth. Their durability was, on the whole, daunting—*Lyctus* beetle adults exposed to dosages of 48,000 rad continued to lay eggs, and mature eggs of furniture beetles withstood exposures between 48,000 and 68,000 rad (Bletchley and Fisher 1957); exposure to 20,000 rad failed to kill adults of the lesser grain beetle or flour beetle (Cornwell et al. 1957). As a point of comparison, exposure to 1,000 rad is generally enough to kill a human.

No one, though, thought to aim his or her cathode ray tubes, cobalt 60 sources, or Van de Graaff generators at cockroaches until Wharton and Wharton (1957) directed 1,000 rad against the American cockroach *Periplaneta americana;* they found that this sublethal dose interfered with production of pheromones, or sex attractants. In a subsequent study (Wharton and Wharton 1959), these same authors conclusively demonstrated that the American cockroach was, compared with the rest of the known irradiated insect world, a wimp; *P. americana* died at doses of 20,000 rad. In comparison, it was noted that the lowest dosage that would kill the entire sample of fruit flies was 64,000 rad. For the parasitic wasp *Habrobracon,* it was 180,000 rad.

In retrospect, it could be argued that the American cockroach might be atypically sensitive to radiation as far as cockroaches go, but subsequent studies on other species have not established any longstanding preeminence for cockroaches among the ranks of the radiation resistant. A subsequent study of the effects of ionizing radiation on *Blattella germanica,* the German cockroach, found that doses as low as 6,400 rad killed 93 percent of nymphs after thirty-five days, and effects on reproductive capacity could be detected at doses as low as 400 rad. Granted, German cockroaches proved capable of surviving ten times the dosages over the same time period that would be lethal to humans, but in point of fact they ultimately succumbed to dosages that don't even disturb many other insect species. So, why is it that Americans came out of the atomic era with the image of a survivor cockroach rather than a survivor fruit fly or a survivor lesser grain borer?

Probably because lesser grain borers and fruit flies don't fit the image of the ultimate survivor. People will continue to believe that cockroaches will survive nuclear war no matter how powerful nuclear weapons become or how large arsenals grow. As David George Gordon (1996) pointed out in his book *The Compleat Cockroach,* today's run-of-the-mill one-megaton thermonuclear devices are at least seventy times more powerful than the fifteen-kiloton bomb dropped on Hiroshima (and even these pale in comparison with the fifty-eight-megaton nuclear device tested by the former Soviet Union in 1961), so, even if a cockroach could have survived Hiroshima's bombing, it wouldn't have much hope of surviving even a nuclear skirmish between rogue states today, much less a battle between nuclear superpowers.

Cockroaches will likely remain in the public conscience as the most radiation resistant of all creatures, all data to the contrary. The bacterium *Deinococcus radiodurans* briefly enjoyed quite a bit of favorable publicity (for a bacterium at least) when its genome was completely sequenced. *D. radiodurans* (as the specific epithet suggests) is without doubt the most radiation-resistant organism known on the planet. A pinkish bacterium that smells vaguely of rotten cabbage, it was originally isolated from canned meat that had spoiled despite being irradiated (it has also turned up in irradiated fish and duck meat, the dung of elephants and llamas, and granite from Antarctica). It grows happily in radioactive waste sites in the presence of levels as high as 1.5 million rad (keep in mind that's over 1,000 times what it takes to kill humans and sterilize American cockroaches). In a frozen state it may even be able to withstand 3 million rad. Notwithstanding its astonishing biological abilities, I don't see a stinky pink bacterium ever displacing the venerable cockroach in the public imagination as the sole survivor of whatever havoc humans wreak on the planet. For that matter, I can't imagine any metal/punk bands in the near future choosing the name *Deinococcus radiodurans* either (although The Lesser Grain Borers or Confused Flour Beetles definitely have potential).

the olympian flea

I DON'T KNOW WHO first came up with the idea of measuring lengths in units of football fields, but I imagine it was an entomologist. Football fields are the preferred units for expressing equivalent distances that insects, particularly fleas, could jump were they the size of men. No sexist intent, here—for some reason, these equivalencies always seem to be measured with men in mind. My personal theory is that only a guy would care if he could outjump a flea if he were the same size as a flea. Football fields are routinely used to illustrate the prodigious athletic capabilities of insects. According to a standard introductory entomology text (Borror, DeLong, and Triplehorn 1976), "When it comes to jumping, many insects put our best Olympic athletes to shame; many grasshoppers can easily jump a distance of 1 meter, which would be comparable to a man broad-jumping the length of a football field." Information in the 1990 *Guinness Book of World Records,* proclaiming *Pulex irritans* the "champion jumper among fleas" reported, "In one American experiment carried out in 1910 a specimen allowed to leap at will performed a long jump of 330 mm (13 in) and a high jump of 197 mm (7.75 in)" (p. 41). These statistics in turn inspired some calculations on the Bugman Bug Trivia Web site: "so, let's do the math . . . after scouring our extensive piles of resources, the best estimate of flea length we could find was 1/16 to 1/8 of an inch. So let's take the large estimate ('cause that's more conservative). 1/8″ is about 3 mm. So, a flea can jump about 110 times its length. Now, for example, if you are 5 feet tall (or long) and could jump 110 times your length, you could jump about 550 feet, which is about 183 yards or, nearly 2 football fields!"

I suppose these analogies are helpful to sports fans, but I have no clear concept of how long a football field is (having attended only one-and-a-half football games in my entire life, both of which took place over thirty years ago). Moreover, "football field" as a unit means different things in different countries. As I understand it, Canadian football is played on a field that's 110 yards long. "Football" in Europe refers to soccer, and that field varies from 100 to 130 yards long. Do European fleas make the conversion when they jump?

Admittedly, not all of the jumping analogies revolve around football. Whereas football field units seem well suited to illustrating the length of a flea's broad jump, they would seem far less useful to illustrate the relative height of a flea jump. Indeed, more often than not, high-jump equivalents are measured in units of buildings, usually relatively famous ones. The utility of such comparisons depends on one's familiarity with scenic landmarks; in an article about the Olympic prowess of animals, R. McNeill Alexander references the apparently popular comparison equating a flea's thirty-centimeter jump to "a man jumping over St. Paul's Cathedral" (Milius 2008), which for American stay-at-homes is unenlightening at best. But the football field as a unit of measure is so firmly entrenched in the popular consciousness that occasionally it serves as a unit of height. At the "Super bugs? Whimpy [sic] humans?" Web site, it is stated that "Fleas can jump over 80 times their own height, the equivalent of a 6 foot tall human jumping over a building 480 feet (more than 1 and a half football fields) high!" Short of a seismic cataclysm, when can people see football fields stacked vertically?

The problem with all of these calculations, of course, is that they fail to take into account the ratio of surface area to volume. Small organisms, such as insects, live in a world dominated by surface forces. The bigger the organism, the greater its volume (which is a function of length times width times height) relative to its surface area (which is a function of length times width). Cubic dimensions scale up faster than do squared dimensions, so, as organisms increase in size, surface area can't keep pace with volume. Muscle strength increases with cross-sectional area, so a small organism (like a flea) has muscles with a relatively high cross-sectional area moving a relatively small volume. The muscles themselves aren't stronger—they're just doing smaller jobs relative to their size. A six-foot flea would have about the same muscle strength as a six-foot man, so in all probability it wouldn't be able to leap over any goalposts unless they were knocked flat and lying on the ground.

In fact, insect muscles might not even be as strong as vertebrate muscles on an absolute basis. As the great twentieth-century biologist J. B. S. Haldane famously wrote in his essay "On Being the Right Size,"

> the height to which an animal can jump is more nearly independent of its size than proportional to it. A flea can jump about two feet, a man about five. To jump a given height, if we neglect the resistance of air, requires an expenditure of energy proportional to the jumper's weight. But if the jumping muscles form a constant fraction of the animal's body, the energy developed per ounce of muscle is independent of the size, provided it can be developed quickly enough in the small animal. As a matter of fact an insect's muscles, although they can contract more quickly than our own, appear to be less efficient; as otherwise a flea or grasshopper could rise six feet into the air. (Haldane 1928)

Although insect muscles may be less efficient, they're still capable of some amazing feats. Some insects have muscles that function in ways unlike any muscles humans have (or any other organism, for that matter). *Odontomachus bauri* is one of a group of ants collectively called trap-jaw ants; these ants are capable of snapping their jaws shut with incredible speed. Using an extremely sophisticated high-speed camera recording at 100,000 frames per second, my colleague Andy Suarez and his collaborators measured, on average, closing speed ranging from 35.5 to 64.3 meters per second and accelerations of 100,000 g (Patek et al. 2006). *Odontomachus bauri* can shut its mouth in less than 100 nanoseconds. These investigators also determined that when the jaws close, they exert a force of 47 to 69 millinewtons, which is approximately 370–500 times their own body weight. The speed of the jaws changes through the arc of closing, with the mandibles slowing down past the midline, possibly to reduce the risk of smashing if they hit each other.

This spectacular mandibular prowess raises the question as to why any organism has a need to snap its jaws shut with such force and speed. These ants are remarkably versatile. They can use their trapjaws to ensnare prey, but they can also, by slapping their jaws against a hard object (such as an intruder) or against the ground, propel themselves into the air. The bouncer defense jumps, launched off an intruder, can reach 40 centimeters horizontally. Escape jumps, launched from the ground, cover shorter distances but greater heights, up to 8 centimeters. Even more impressive than the distances covered, though, is the fashion in which they're covered. They don't just slap their mandibles against a surface; a stereotyped set of behaviors sends the ant spinning head over all six heels, with a spin rate that can peak at more than 60 revolutions per second.

One wonders what football analogy can be used to place that feat in human terms. The world record for "fastest spin on ice skates," set by Natalia Kanounnikova at Rockefeller Center in New York City, is 308 revolutions per minute (rpm). During jumps, ice skaters can reach 420 rpm, or about 7 revolutions per second. But that's about one-ninth the spin rate of a trapjaw ant. Football players don't routinely spin, at least by design, but in terms of spinning things on a football field, even the football doesn't measure up to a trapjaw ant. Typically, a tossed football manages about 8–10 revolutions per second, with an acceleration of about 8 meters per second. So, the next time a football player is bragging about his physical prowess, maybe a comparison with the trapjaw ant will shut him up—but even if it does, it will likely take longer than 100 nanoseconds.

the prognosticating woollyworm

INSECTS HAVE SOME TRULY spectacular abilities. There are species that can walk on ceilings, for example, or chew through lead cable, or fly through the air at speeds exceeding thirty miles per hour, so I guess it's not surprising that people occasionally believe them to have supernatural powers. There's the pervasive notion that certain caterpillars, which are otherwise not spectacularly well-endowed in the brain department, are capable of predicting winter weather. I'm not sure where or how this bit of folk wisdom arose, but it is certainly firmly entrenched in the popular psyche.

As the story goes, the larval stages of moths in the family Arctiidae, caterpillars called woollybears or woollyworms, provide an indication of the severity of the upcoming winter by virtue of the width of a central orange or red band across the middle of their otherwise black bodies. In theory, the broader the band, the milder the winter. This prognosticative ability is not shared by all woollybears—according to custom, it's only the caterpillars of *Isia isabella,* the banded woollybear (which grows up to be the far less charismatic beige-winged Isabella tiger moth). This raises some confusion. One woman, who came across the uniformly white caterpillars of *Diacrisia virginica,* the Virginia tiger moth, sent me an email message expressing her concern that the total lack of dark color meant we might be in for another Ice Age.

How this connection with winter arose is uncertain, but the variation in band width has intrigued entomologists for decades. Frank Lutz (1914) described experiments with humidity, concluding on somewhat tenuous grounds that the width of the band was a function of the humidity experienced by the larvae while they grew; "In the course of some work at the Carnegie Station for Experimental Evolution I found that I could change to a surprising extent the markings on the larva of a moth *(Isia isabella)* by varying the temperature at which they fed and moulted. However, such changes were much more definite when the temperature was kept constant and humidity varied. I did not have the necessary apparatus for getting accurate control of either factor, but I feel confident that temperature had little or no direct influence. It was acting through its influence upon humidity."

Probably the last attempt to investigate the scientific basis of the meteorological predictive powers of the banded woollybear was launched by Charles Curran of the American Museum of Natural History. He conducted a series of experiments beginning in 1947, attempting to correlate band width with winter severity, but abandoned this work in 1955. He concluded that the correlation was predictive about half the time (making woollybears about as successful in predicting winter weather as contemporary meteorologists). Maybe the most reasonable explanation for the variation is that banded woollybears start out life with a broad band that narrows as they approach the larval stage in which they'll pass the winter. A mild fall ostensibly allows them to develop faster than a cold fall, which means they'll enter their overwintering dormant state, or diapause, with less black hair than their more frigid counterparts.

The ability to predict the weather has been ascribed to a wide variety of arthropods on a variety on continents, but most reports of arthropod meteorological forecasting are sufficiently vague as to instill doubt that any real biological phenomenon is responsible; after all, in many places in the world human meteorologists content themselves with predicting probabilities of rain. Far more impressive, however, is the ability of at least one species of cricket to measure actual temperatures. Amos E. Dolbear was an American physicist who is regarded as the undisputed inventor of the electric gyroscope, the opeidoscope (a device for visualizing sounds), the wireless telegraph, and a novel form of incandescent lighting. He was also the disputed inventor of the telephone. Although he devised a magnetic telephone receiver in 1865, over a decade before Alexander Graham Bell did much the same thing, his failure to file a patent ultimately led the Supreme Court to decide the lawsuit *Dolbear et al. v. American Bell Telephone* in favor of Bell. It's a strange quirk of history that he is best known as the eponymous source of Dolbear's law.

In 1897, for reasons apparently lost to history, Dolbear took time from his busy schedule of inventing to publish a two-page paper in the journal *American Naturalist* (1897) noting that although "an individual cricket chirps with no great regularity when by himself . . . [a]t night, when great numbers are chirping the regularity is astonishing." Moreover, he pointed out that the "rate of chirp seems to be entirely determined by temperature and this to such a degree that one may easily compute the temperature when the number of chirps per minute is known." The paper concluded with a mathematical equation to convert number of cricket chirps (N) into temperature (T): $T = (N - 40)/4$.

This mathematical equation became widely cited as Dolbear's law, but many of the citations derived from the fact that Dolbear, as a physicist, neglected to identify which species of cricket he was calibrating. It turns out that Dolbear's law doesn't apply to all crickets. The general consensus over time is that Dolbear's mathematical equation was empirically derived from one particular species, the snowy tree cricket, or *Oecanthus niveus*. Moreover, subsequent studies, beginning only a year after Dolbear's publication, determined that the law isn't exactly unbreakable; chirp rates of the snowy tree cricket are affected by wind currents, physiological condition, and genetic background, among other things (Frings and Frings 1957). And by 1899, Robert T. Edes published a note in the *American Naturalist* pointing out that "A few years ago a note appeared in the *Boston Transcript* calling attention to the very exact dependence of the rapidity of the chirps upon the temperature of the surrounding atmosphere and giving a formula therefore . . . possibly the same" as the one provided by Dolbear. Neither Dolbear nor Edes cited the paper by Margarette W. Brooke, published in *Popular Science Monthly* in 1881, titled "Influence of temperature on the chirp of the cricket." In it she reported the results of her test of a theory by "a writer on the *Salem Gazette*, signing himself W. G. B." who provided a "rule for estimating the temperature of the air by the number of chirps made by the crickets per minute: 'Take seventy-two as the number of strokes per minute at 60° temperature, and for every four strokes more add 1° and for every four strokes less deduct the same.'" Her test revealed a "remarkable accordance," which raises the question as to why this relationship is not now known as "W. G. B.'s law."

There are dozens of variants of Dolbear's law, including variants that require counting the number of chirps in fourteen seconds and adding forty-two, or, for those in a real hurry, the number of chirps in seven seconds and adding forty-six (Clausen 1954). But for those with too much time on their hands, there's an even more labor-intensive way to figure out the temperature —the Ramsey ECS1, otherwise known as the Electronic Cricket Temperature Sensor Kit. For less than twenty-five dollars, anyone who doesn't have a thermometer but does have a credit card and an Internet connection can build from scratch a digital cricket that changes chirp rate with temperature ("Just count the number of chirps over a 15 second interval, add 40, and you have the temperature in degrees Fahrenheit!"). It runs on a nine-volt battery, which isn't included, and "if it drives you nuts, you probably can squish it under your feet to make it stop . . . but that voids the warranty!" I can't help thinking that, had he lived to see the electronic cricket sensor, the inventor of the opeidoscope probably would have approved.

the queen bee

QUEEN BEES have been getting a lot of press of late and not much of it is especially positive. In 2002, for example, Rosalind Wiseman wrote a book titled, *Queen Bees and Wannabes: Helping Your Daughter Survive Cliques, Gossip, Boyfriends, and Other Realities of Adolescence.* The thesis of the book is that some adolescent girls possess "evil popularity" and use it ruthlessly to their advantage to disenfranchise or shut out other girls who have in some way incurred their wrath. The implication of the titular metaphor is that the teen queen has absolute power over her subjects (Wiseman 2002), presumably of the sort that western honey bee queens exert over the 30,000 to 50,000 workers in the typical *Apis mellifera* colony.

It's not the first time queen bees have come to the metaphorical rescue of psychologists. In 1973, the term "queen bee syndrome" (Staines et al. 1973) was coined to describe women who achieved success in a predominantly male work force by turning against other women, a negative stereotype that has not been validated (Mavin 2008). Notwithstanding the absence of evidence that such a syndrome exists, the metaphor persists. The radio pundit Rush Limbaugh freely applied the metaphor to political analysis in describing the politician Nancy Pelosi: "Do you know what the Queen Bee Syndrome is? There will not be two women sharing power. One of the women will see to it that the other woman is under the bus. So Nancy Pelosi today, holding her weekly press conference, was asked about the concept of a "dream ticket" for the Democratic Party. Hillary and Obama, or Obama and Hillary. The Queen Bee in Washington, Nancy Pelosi, threw cold water on this whole idea. She said, 'Take it from me—that won't be the ticket'" (Limbaugh 2008).

It's indeed the case that each honey bee colony has but a single queen and that newly emerged queens do traverse the hive and sting to death any other presumptive queens unfortunate enough to have taken a little too much time to develop. Otherwise, the analogy doesn't hold up very well at all. As a symbol of absolute power, the queen honey bee falls a bit short. Once she establishes herself in the hive and sets off to mate, she is doomed to an existence of endless egg-laying, at a rate of about sixty per hour. She is continuously tended by a retinue of workers who ply her with food, groom her meticulously, and otherwise push and position her to meet the needs of the hive. Not only does she have absolutely no privacy, were her retinue to abandon her she'd most likely die for want of any capacity to fend for herself. At least the workers get to fly out of the hive on occasion and check out the scenery—the queen doesn't see much beyond empty wax cells awaiting her eggs. Any breakdown in the production line—by virtue of exhaustion, boredom, or old age—invites a process called supersedure, whereby workers raise a new queen and then dispatch the old one by clustering around her, generating body heat, and cooking her to death.

I guess it's a sign of progress that at least some bee-related metaphors—the one-queen-per-hive concept, for example—have some relationship to honey bee biology. Such wasn't always the case. The concept of male domination over females is so deeply ingrained in western culture that centuries passed during which, despite all evidence to the contrary, male scientists insisted that honey bee colonies are ruled by males, as were most human societies. To think otherwise, according to Prete (1991), necessitated challenging "the very idea of an orderly universe" and, starting in the sixteenth century, authors of scholarly beekeeping texts had to go through extraordinary contortions to ignore gathering evidence of the queen's femininity. In his 1607 *History of Four-Footed Beasts,* Edward Topsell, like earlier writers, reported that male bees lack stingers and do no apparent work in the hive; to reconcile the facts with his desire to hold up bees as a model for idealized (British) society, his tortuous explanation was that

> The prince of *philosophers* confoundeth the sexe of Bees, but the greatest company of learned Writers do distinguish them: whereof they make the feminine sort to be the greater. Others again will have them the lesser with a sting: but the sounder sort (in my judgment) will neither know nor acknowledge any other males but their Dukes and Princes, who are more able & handsome, greater and stronger than any of the rest, who stay ever at home . . . as those whom nature

pointed out to be the fittest to be standard-bearers . . . and ever to be ready at the elbows of their loves to do them right . . . If any Souldier looseth his sting in fight, like one that had his Sword or Spear taken from him, he is presently discouraged and dispaireth, not living long, through extreamity of griefe. Bees are governed and doe live under a Monarchy . . . admitting and receiving their King . . . by respective advise, considerate judgement, and a prudent election.

Charles Butler's careful dissections and masterful account of honey bee biology in his 1612 book *The Feminine Monarchie* should have put the matter to rest once and for all ("But heer' is bot' Reason and Sens consenting, doo plainly proov' . . . dat bot' de Princ' and hir armed subjects are Shees . . . Bees or breeders as deir leaders: and again, Bee's . . . ar femal's") but many contemporaries were reluctant to abandon their idealized conceptions. The Reverend Samuel Purchas wrote *A Theatre of Politicall Flying-Insects* in 1657 as a handbook both for beekeeping and clean living; included in the 300 sermons were many references to the life of bees. Although he admitted that "Though a king in place and power . . . [the monarch] is in sex a female," he nonetheless refers to the queen as "he": "Bees . . . [live] under one commander who is not an elected Governor . . . nor hath hee his power by lot . . . nor is hee by hereditary succession placed in the throne . . . but by Nature hath bee the SOVEREIGNTY over all, excelling all in goodliness and goodness, and mildness, and majesty" and even went so far as to suggest that the "queen" "injects a spermatical substance thick like cream" into the wax cells in which future queens are developing (Purchas 1657).

But even societies that recognize the true gender of the queen bee don't quite understand how honey bee societies are structured. In reality, the queen bee is the epitome of the traditional female—barefoot (times six) and pregnant. The queen bee's only job is to lay eggs, and this is what she does, twenty-four hours a day, seven days a week, at a rate of over 2,000 per day. In fact, she can lay more than her own weight in eggs in a day. It's true that the queen is tended by a retinue of workers who feed, groom, and protect her and who even cart her waste out of the hive. But she's not in any position to wield power or influence over any of them; she's in fact more or less at their mercy.

Gender confusion with respect to bees extends well beyond the queen. Probably even less well understood than the status of queen bees is the status of male bees. In cartoons and advertising, bees are almost invariably depicted as male. Donald Duck faced off in a dozen cartoon episodes against Spike the Bee, and Jerry Seinfeld voices Barry B. Benson, a male bee who in the feature film *The Bee Movie* announces his intention to sue the residents of New York for theft of his honey. Spokesbees ranging from the Wheat Honey's Buffalo Bee from the 1950s to the Honey Nut Cheerios Bee of the present day are unmistakably male. Although they don't sport facial hair, their voices are male and they dress like guys (even down to the pointy cowboy boots). Antonio Banderas, about as masculine as any man alive today, lends his voice to the spokesbee for Nasonex, a preparation for treating allergies.

The irony in all of these depictions is that male bees have nothing to do with honey except to eat it when it is handed to them by a female worker. A male bee has nothing to do with carrying pollen around, either. They're not called "drones' without reason. Drones are incapable of foraging for pollen or nectar, caring for offspring, or, indeed, even caring for themselves. All they can do is inseminate the queen, and once that's done they die (by virtue of the fact that their genitalia, firmly lodged in the queen's bursa copulatrix, tear off once the act is complete, leaving them to fly away missing many essential internal organs and probably in no mood to sing the praises of breakfast cereals).

Beginning in October, 2006, billions of bees began disappearing without a trace, ostensibly due to a mysterious condition called "colony collapse disorder." Theories proliferated wildly and, by virtue of the fact that I had written an opinion piece for the *New York Times* about the phenomenon, many people shared their own theories with me. In May, 2007, a pet psychic emailed me to explain that she had been able to communicate directly with two bees. She asked them what was going on and was told by one bee that his usually trustworthy navigation system had failed him. I don't think that she was making up the story, but there was a problem nonetheless inasmuch as her informants were identified as male. Male bees have nothing do to with collecting pollen or nectar and do not navigate at all (except for pursuing virgin queens). Once they mate and lose their genitalia by explosive force, they have no need to return home because they die almost immediately thereafter. So, either her interpretation skills need honing, or these drones were total poseurs.

the right-handed ant

WHENEVER I HAVE A QUESTION about ants, rather than consult the literature or check the Internet I ask my colleague Andy Suarez, who, it seems, knows just about everything there is that's worth knowing about ants. So I was surprised that he didn't know anything about what was billed as one of "the Most Interesting and Unusual Facts on the Net." As an entomologist, I feel no responsibility for checking the veracity of facts that relate to any organisms with fewer than six legs, so I'm willing for the sake of argument to believe that, as this Web site declaims, all polar bears really are left-handed and all porcupines really can float in water (and have 30,000 quills on their bodies, which are replaced every year). But I had serious reservations about an entomological "Interesting and Unusual Fact"—namely, that an ant "always falls over on its right side when intoxicated."

Suspecting there was a body of literature involving ant intoxication of which I was blissfully unaware, I asked Andy what he knew. Andy, who has probably read every scientific publication on ants ever written, had never heard that intoxicated ants always fall on their right side. Moreover, he raised yet another puzzling question. Since he works on Argentine ants in Argentina, he wondered if the right-side rule would apply in the Southern Hemisphere, or whether, like clockwise toilets that reverse directions when they cross the equator, South American ants fall on their left side after, say, imbibing too much pulque.

To find out just how unusual this unusual fact was, I did a quick search of the Internet and found that in cyberspace this fact is well established. It appears, for example, on the site Answerbag (along with the observation that deer have no gall bladders), and it also appears on the Stuff You Didn't Know site, which, in addition to the left-handed polar bears, also provided the helpful information that "it takes four hours to hardboil an ostrich egg". I also found it on Yahoo! Answers, on a site called Unsolved Mysteries under the heading "A multitude of weird things that you probably didn't know," along with the floating porcupines and the statement that "a duck's quack doesn't echo." And it appears under the heading "Worthless Information" on a site run by one Michael A. Urich of LaPorte, Texas, sandwiched between "A crocodile cannot stick its tongue out," and the ubiquitous left-handed polar bears.

Many of the sites providing this information about ants also included snarky remarks about what motivated members of the scientific establishment to spend their time getting ants drunk.

One Shyamala Ramanathan was inspired to ask on her site "Inspired to Blog," "How DOES one get an ant intoxicated? On what? How much of whatever liquor does it take? Does it prefer beer to spirits? What sort of a glass does it use? Does it prefer a straw? How does one keep an ant merely (and possibly merrily) intoxicated without it going over into the blind raving drunk zone? Would the ant prefer a bar or a pub, or would it be pleased to spend Happy Hour in a lab, in a spirit of scientific endeavour?"

I must confess, my mind was fairly boggled along the same lines (and I was already distracted by recurring mental images of floating porcupines). A search of the refereed scientific literature revealed a sizable body of literature on intoxicated arthropods of all descriptions. What appears to have been the first scientific effort to deliberately intoxicate an ant dates back to 1878 and involved no less august a scientist than Sir John Lubbock, First Baron Avebury, banker, politician, archeologist, and naturalist. I found an account of his experiments with ant intoxication in his book *Ants, Bees, and Wasps* (1882), which I reflexively bought in a used book store years ago, hoping that someday I might actually have occasion to read up on Victorian experiments in Hymenoptera biology.

Chapter 5, titled "Behaviour to Relations," concludes with an account of a series of studies examining how ants behave toward "friends and strangers" after various degrees of impairment. Lubbock's overall objective was "to ascertain whether ants knew their fellows by any sign or pass word," in particular "to see if they could recognize them when in a state of sensibility." He noted that ants that had been chloroformed were picked up and tossed into a moat of water irrespective of whether they were friends or strangers. Upon realizing that "the ants being to all intents and purposes dead, we could not expect that any differences would be made between friends and strangers," Lubbock decided that intoxicating the ants would be a better test of nestmate recognition. Despite the difficulties in obtaining "the requisite degree of intoxication," he noted that "the sober ants seemed somewhat puzzled at finding their intoxicated fellow creatures in such a disgraceful condition, took them up, and carried them about for a time in a somewhat aimless manner." Disgraceful condition notwithstanding, "out of forty-one intoxicated friends," thirty-two were ultimately escorted home by their long-suffering nestmates (and only nine "thrown into the water").

Lubbock's keen observations, however, included no mention of directional staggering. Nonetheless, the work attained iconic status and is cited to this day in the scientific literature, although somewhere along the way the tale was embroidered for greater effect. In 1907, according to the account of the experiments by John Holmes Agnew and Walter Hilliard Bidwell for *Eclectic Magazine,* the ants were not only intoxicated, they were "reeking of whisky."

Lubbock would surely have been interested in the work of Charles Abramson, who may have been the first to investigate the behavior of the intoxicated honey bee, *Apis mellifera.* In proposing the honey bee as a model for understanding effects of alcohol, Abramson and his colleagues (2000, 2007) noted that "consumption of 10% and 20% ethanol solutions decreases locomotion . . . [and] ethanol solutions greater than 5% significantly impair Pavlovian conditioning of proboscis extension" (Abramson et al. 2000). So there is definitive evidence that, whatever other social skills they possess, honey bees just can't hold their liquor. Further studies showed that alcohol's effects on bees include "self-administration, disruption of learning and locomotion when traveling home [to the hive], preferences for commercially available alcoholic beverages," and an increase in aggressive behavior in Africanized bees (Abramson et al. 2005). Intoxicated bees also don't communicate very well; Bozic et al. (2006) demonstrated that intoxicated bees have problems managing the intricacies of the waggle dance.

There are limits to the utility of the honey bee as a model organism for investigating human responses to alcohol. I don't know how an experiment might be designed, for example, to determine if bees that are intoxicated think their jokes are way funnier than they actually are. But they're surely better models than a bean plant or a marine isopod, both of which have been intoxicated in the name of science. Noting that ethanol had a clear and repeatable effect on the periodicity of leaf-movement rhythm of *Phaseolus* bean plants, Enright (1971) leaped across a gaping taxonomic divide to see if it similarly affected the "free-running tidal rhythmicity of the sand-beach isopod, *Excirolana chiltoni*." Aside from inducing some aberrant behaviors, which included burying themselves in sand with their abdomens, "ostrich-like," projecting above the sand, and disturbingly high mortality after the first twenty-four hours, Enright succeeded in demonstrating that ethanol does in fact increase the length of the free-running rhythmic tidal activity of the isopods. Whether ethanol affects rhythmicity by the same mechanism in bean plants and sand-beach isopods was left as unresolved, but he allowed that "the present phenomenon is probably not directly involved in the subjective experience that alcoholic beverages make the time pass faster."

I'm not sure I fully accept the idea that sand-beach isopods and bean plants have reactions to alcohol identical to reactions of human consumers of alcohol; I don't know, for example, how a bean plant can be loud and obnoxious at a party. But humans, arthropods, and alcohol do seem to be inextricably linked culturally. Mescal, for example, an alcoholic beverage distilled from agave, is traditionally bottled with a "gusano de maguey," a small white caterpillar (often *Aegiale hesperiaris*), floating near the bottom (a tradition, by the way, that dates back not to the ancient Aztecs but to 1950, when a Mexican entrepreneur, Jacobo Lozano Paez, dreamed up the idea as a way of authenticating the agave plant origin of the product). Today consumption of large quantities of mescal cause college students not to fall on their right side but rather to eat the pickled insect floating in the bottom of the bottle. A query on YouTube with "tequila worm" yields 167 videos, despite the fact that, technically speaking, bottles of mescal, and not tequila, are the ones with the "worms."

There are also several brands of vodka featuring a pickled scorpion floating in the bottom of the bottle. In one brand putatively from Thailand, the scorpion is a "farm-raised" *Heterometrus spinifer*. Skorppio, a vodka imported from England, also comes with a pickled scorpion, unidentified but also "farm-raised." The scorpions, interestingly, are, according to the label, "subject to analysis certified by the Chamber of Commerce of Pismo Beach, CA, U.S.A., to confirm that no harmful substances are present." I had no idea that this sort of thing has been going on in Pismo Beach. Until now, the only cultural reference point I've had for the place is that it was Bugs Bunny's destination when he didn't turn left at Albuquerque in the Warner Brothers classic cartoon *Ali Baba Bunny*. I wonder if he would have made it to Pismo Beach by turning right in Argentina.

the §ex-enhancing spanishfly

SOMEDAY, IF I GO MISSING and investigators try to figure out what has happened by identifying Web sites I've visited, I may have a lot of explaining to do. The World Wide Web has a habit of taking me places I didn't ever intend to go. A few years ago, for example, I wanted to look up some information on the biology of what was at the time an emerging pestiferous species—*Harmonia axyridis,* a nonnative species of coccinellid beetle. This insect had been imported for the biological control of arboreal aphids, but it never really rose to that particular challenge; however, about two decades after the initial introductions, *H.*

axyridis became noxious by virtue of its habit of overwintering in aggregations numbering in the tens of thousands inside people's homes (Koch 2003). I thought I'd search using the common name of the insect, but I had a problem—there are at least two common names. Although *H. axyridis* is known as the multicolored Asian lady beetle, it is also known colloquially as the multicolored Asian ladybug. To capture as many sites as possible, I decided to search just "multicolored Asian lady." That turned out to be a major mistake. Some auditor someday will ask me why I checked out the "Sexy Beautiful for Dating" site on my office computer and probably won't believe my story.

There are times, though, that entomology and not convergent orthography takes me to risqué sites. In particular, I periodically check up on *Lytta vesicatoria,* the notorious Spanishfly. This, of course, is the meloid beetle (not a fly) that has been used (or abused) for centuries as an aphrodisiac throughout Europe (and not just Spain). The clearly imprecise term "Spanishfly" dates back to the seventeenth century, before insect taxonomy established strict rules for ordinal membership. At that time, these insects were highly prized for their medicinal value. The blood, or hemolymph, of *Lytta vesicatoria* causes blistering and engorgement of mucous membranes due to its abundant supplies of the terpene anhydride cantharidin, a genuinely pharmacologically active substance. Among the bodily frailties for which Cantharides, as they were known medicinally, were prescribed since the days of Hippocrates included but were not limited to dropsy, rheumatism, carbuncles, leprosy, and gout. Given the relative rarity of legitimately active ingredients in medicines of the era, it's not surprising that Spanishfly was so highly regarded; other prescriptions of the era included such inert yet off-putting components as lizard dung and jackal bile.

The reputation of Spanishfly as an aphrodisiac stems from the fact that engorgement of mucous membranes with subsequent inflammation and itchiness was regarded by some as a desirable state. Historical figures running the gamut from Ferdinand the Catholic to the Marquis du Sade were said to have made use of the stuff for noble and not-so-noble purposes (Karras et al. 1996). As an aphrodisiac, though, Spanishfly left a lot to be desired; as little as 30 mg could actually cause horrible, painful, and potentially embarrassing death. Thomas Muffet, in his seventeenth-century classic, *Theater of Insects,* recounts in the chapter on Spanishfly the story of "a certain married man . . . fearing that his stopple was too weak to drive forth his wife's chastity the first night, consulted one of the chief Physicians, who was most famous, that he might have some stiffe prevalent Medicament, whereby he might the sooner dispatch his journey. But when it was daybreak almost, there followed a continual distending of the yard without any venereous desires, and after that bloudy urine, with inflammation of the bladder, and the new married man almost fainted away." The man probably didn't realize how lucky he was—Muffet also recounted the story of the unfortunate "Noble Man of Frankfort," who was given Cantharides by a physician to cure a nasty case of dropsy. Unfortunately, the medicine "killed him with lamentable torments."

Today, the Food and Drug Administration restricts the medicinal use of cantharidin to warts and a few other skin conditions, but even so, effective relief for erectile dysfunction is obtainable with the click of a mouse, it would seem; in fact, offers for such products appear in my email inbox almost as frequently as do queries from wealthy Nigerian widows seeking urgent business relationships. Although it would be reasonable to think that there might be little demand for aphrodisiacs derived from animal parts or secretions that can kill you, Spanishfly is alive and well on the Internet. The site "Spanish fly aphrodisiac" boasts of formulas for both men and women. For men, there's Kriptonite, which the site claims can "Boost the ability to increase the Inner Sexual Conscience of the mind, enhancing your confidence and desire to achieve various sexual advances." Although the standard formulation is only $69.95, Kriptonite X 12 sells for $503.64 per bottle; this may seem pricey, but for the economically minded there's a 40 percent discount for orders of a dozen bottles, although what anyone would do with a dozen bottles exceeds my capacity to imagine. And for women, the site markets a liquid "originally extracted from beetles that live in Spain. The male beetles use this chemical to sexually seduce females into having sex with them. The chemical has now been reproduced in the laboratory at highly concentrated levels. It increases sexual stimulus, it improves the disposition towards sexual activity and improves mood." In an admirable example of truth in advertising, the site states, "The substance Spanish Fly liquid irritates the urogenital tract and produces an itching sensation in sensitive membranes, a feeling that allegedly increases a woman's desire for intercourse." That "allegedly" is certainly well placed., inasmuch as I can't imagine to what alternate universe this cause-effect scenario might apply.

A predictable consequence of looking for love in all the wrong places is that Spanishfly continues to let men down. In 1954, one "Mr. X," an employee of a pharmacy, offered coconut candy to two female clerks, both of whom within hours began vomiting large quantities of blood and were dead by the next day. "Mr. X" had stolen the cantharidin from the druggist under the pretext of needing it for a neighbor's underperforming rabbit and laced the coconut treat with it in the hope of winning the amorous attention of his hitherto uninterested co-workers. He served five years in jail for manslaughter. Nine years earlier, a dentist who had given another man 1 gram of cantharidin as an aphrodisiac, which killed him, was only sentenced to thirty days and fined 7 Danish kroner (Nickolls and Teare 1954). As recently as 1996, four patients arrived at an emergency room in Philadelphia presenting with a range of symptoms that included genitourinary hemorrhage, endstream dysuria, disseminated intravascular coagulation, and vomiting of "pink-tinged fluid"; this constellation of symptoms was brought on by the addition ten hours earlier of several drops of Spanishfly to an orange-flavored drink at a party (Karras et al. 1996).

Aphrodisiacs are dangerous for reasons other than unwelcome disseminated intravascular coagulation and pink-tinged vomit, as Wang Zhendong, chairman of the Board of the Yingkou Donghua Trading (Group) Co. in Liaoning Province, China discovered. Wang was the brains behind an ant aphrodisiac pyramid scheme. *Polyrhachis vicina* is a Chinese ant that is reputed to possess a variety of medicinal attributes. It's the principal ingredient, for example, in "Hot Rod for Men," an "ancient Chinese holistic formula" that offers, among other things, "incredible sexual stamina" and "yang jing." I don't know what "yang jing" means, but I'm guessing based on context.

Recognizing aphrodisiacs as a growth industry, Wang Zhendong, Donghua Zoology Culturing Co., Ltd., and Donghua Spirit Co., Ltd. between 2002 and 2005 began recruiting investors to raise the potent ants for use in health tonics. By 2005, over 10,000 investors had signed up. The company, though, kept delaying dividend payments. Wang was taken to court and eventually convicted of bilking investors out of more than 3 billion yuan (about $417 million U.S. dollars) and sentenced to death; his death sentence was upheld by the Liaoning Provincial Higher People's Court in February 2008.

Amazingly, a second phony aphrodisiac ant investment scheme had been cooked up in Liaoning by another man named Wang.

Wang Fengyou of the Shenyang Yilishen group promised investors huge profits for raising ants. Every deposit of 10,000 yuan (about $1,300 at the time) was supposed to pay a dividend of 3,250 yuan. The offer attracted thousands of investors in the economically hard-hit province. After repeated failures to pay dividends, the company declared bankruptcy in November 2007, whereupon thousands of angry investors took to the streets across Liaoning Province, protesting government inaction and clashing with antiriot police. Some, having lost their life savings, committed suicide.

So insect aphrodisiacs continue to claim lives, albeit not always as a consequence of their biological activity. Despite the serious consequences of perpetuating the notion that the Class Insecta can offer instant sexual virtuosity, journalists just can't resist the temptation to resort to double entendres to report their story. Thus, a Reuters story from December 14, 2007, reports that "Demand softens for ant aphrodisiacs."

the toilet spider

URBAN LEGENDS are those plausible yet unverifiable stories, generally attributed to a source far removed from the storyteller, that have a bizarre or horrific twist and impart some sort of cautionary lesson. The folklorist Jan Brunvand, a noted authority on urban legends, classifies these stories into a variety of genres; most if not all arthropod urban legends fit comfortably within the genre of contamination stories. Probably the granddaddy of all arthropod urban legends is the "spider in the hairdo" story. In brief, as described in an *Esquire* article on "teenage folklore from the fifties": "A girl managed to wrap her hair into a perfect beehive. Proud of her accomplishment, she began spraying it and spraying it, never bothering to wash it again. Bugs began to live in her hair. After about six months, they ate through to her brain and killed her" (quoted in Brunwand 1981).

This story has changed slightly throughout the years to keep up with the times; the "beehive hairdo" in more contemporary versions becomes dreadlocks, for example, but the essential elements remain the same. Apparently, they've remained more or less the same for over 800 years; a thirteenth-century exempla (a tale used to convey a moral lesson in a church sermon) relates the same fate befell a "certain lady of Eynesham, in Oxfordshire," who "took so long over the adornment of her hair that she used to arrive at church barely before the end of Mass" until "the devil descended upon her head in the form of a spider, gripping with its legs" (Brunvand 1981).

Other urban legends involving spiders include "the spider bite," an account of a woman sunbathing on the beach who brushes away an "insect" crawling along her jawbone and falls asleep, forgetting about the arthropod encounter. A week later, she notices a blister-like growth which, when exposed to the heat of a hair dryer, erupts and produces "hundreds of tiny white baby spiders and pus pouring out of the wound!" In one version of this story, the traumatized woman ends up in a psychiatric ward of the very same hospital where she went to have her boil examined.

Although urban legends are supposed to have moral lessons, I'm not sure what the moral lesson is here—don't fall asleep while you're sunbathing or you'll hatch spiders and go insane? Maybe the lesson is for the spider, who should know better than to lay eggs in human flesh, given that no spider is biologically equipped to live parasitically inside the body of another organism. Still another variant is the "spider in the cactus" tale, which relates the story of a family that receives a cactus as a gift, only to find that it begins pulsating oddly. A call to a nurseryman produces a frantic warning to get it out of the house; once outside, it explodes, producing thousands of baby spiders. According to Brunvand (1993), this story surfaced in Scandinavia in the 1970s and enjoyed a rebirth in the 1990s when southwestern décor became popular again.

Probably the most notorious spider-based urban legend, "the spider in the toilet," isn't an urban legend at all—it was a hoax. Hoaxes differ from urban legends in that they are deliberately created and disseminated. These authors recount the story circulated around the Internet in late 1999 and early 2000 that South American blush spiders (so-called *arachnius gluteus*) were infesting toilet seats in a Chicago-area airport and biting unsuspecting passengers relieving themselves between flights, causing chills, fever, vomiting, paralysis, and, in three cases, death. The story originated with Steve Heard, who concocted and circulated the story in part as an experiment to determine how gullible people could be. Two entomologists at the University of California–Riverside, Richard Vetter and Kirk Visscher (2004), pointed out that the hoax succeeded in part because rampant arachnophobia predisposes some people to believe the worst about spiders. I can't help thinking that the miserable experience of air travel is a predisposing factor as well: "Honey, you'll never believe it—not only was the flight delayed for five hours, but there was no inflight meal service, I missed my connection, the airline lost the luggage, and I was bitten on my right buttock by a lethal South American blush spider in the ladies' room in O'Hare."

Before the widespread adoption of indoor plumbing, a fear of spider bites on one's private parts wasn't necessarily irrational. In the first half of the twentieth century, about 90 percent of reported victims of bites inflicted by the black widow spider were male, and approximately half of those were bitten on their private parts while using an outdoor privy, a preferred habitat for the species. Incidents of such bites dropped precipitously once outdoor privies were replaced by indoor porcelain. This is not to say, though, that airport toilets are necessarily devoid of arthropods. A few years ago, my colleague and longtime collaborator Arthur Zangerl had occasion to take a seven-week trip throughout Europe in search of wild parsnips and parsnip webworms. Both species are native to Europe, so, after studying the interaction throughout select portions of the United States for over twenty years, we decided to expand our horizons and investigate the interaction in the place where both species originated. Art brought back hundreds of samples, dozens of digital images, and many new insights on the geographic mosaic theory of coevolution. Thus, understandably but nonetheless distressingly, it wasn't until months after his return that he got around to telling me about the urinals at Schiphol Airport in Amsterdam.

Art evidently had occasion to use a urinal while at Schiphol Airport and noticed that each urinal has a lifelike etching of a fly located near the drain and just slightly to the left of center. He mentioned this remarkable fact to me because he knows of my interest in cultural entomology, no matter where these interests might lead. I've only been to Amsterdam once in my life. It was twenty-eight years ago, and I arrived by train, not by plane. It particularly bothered me to think that, had I been the one to go to Europe this past summer to look for webworms and passed through the Schiphol Airport, I would never have encountered this interesting cultural phenomenon.

Rather than bemoan my lack of Y chromosomes, I instead went to the Internet. In short order I found photographs of the flies of Schiphol Airport at a Web site not, as you might think, about urinals; rather, it's a site on user interface design. The fly image is cited as a fine example of the use of an icon to assist a user in adapting to a particular technology—in this case, by improving user aim. Apparently, it works; according to an article in the *Wall Street Journal,* etched flies in urinals "reduce spillage by 80%" (Newman 1997). The article, subtitled "Using Flies to Help Fliers," also mentioned the fact that a U.S. subsidiary of NV Luchthaven Schiphol, the company that manages the Amsterdam airport, had just negotiated a thirty-year lease to manage John F. Kennedy Airport in New York. Along with constructing a new $1 billion building on the site, Schiphol USA also plans to etch flies in the urinals at Kennedy. Because of my Y chromosome problem, I can't tell you if that's happened yet. Maybe if our grant is renewed, Art can return to Europe via JFK Airport instead of O'Hare Airport in Chicago and let me know.

Call it what you will, the use of a life-size image of a fly to draw attention and modify behavior has a long and illustrious history. The tradition of trompe l'oeil ("fool the eye") painting, i.e., creating images so realistic that the viewer mistakes them for the authentic item, dates back to the painters of classical Greece. It reached a pinnacle of sorts when the discovery of perspective in the fifteenth century and the invention of optics allowed painters to create images with greater precision in the seventeenth century. The goals of those artists were, of course, quite different from those of the designers of Schiphol urinals, providing a way for artists to showcase their technical skills. Interestingly, trompe l'oeil flies were quite popular among seventeenth-century still-life painters in the Netherlands.

The first application of entomological trompe l'oeil to toilet technology, however, appears to have been British. Victorian urinals sported a variety of images as targets, including literal targets, as on an archery range. Among the animal icons used were honey bees, which raises the possibility that the otherwise repressed Victorians may have been having some fun with the Latin name of the genus, *Apis* (pronounced "A-piss").

I suppose I shouldn't feel too bad about missing out on a gender-biased experience in cultural entomology—it can happen to real insects, too. It's a well-established fact that, where one fly settles, others will be attracted and settle, too. In fact, this so-called flycatcher effect is the reason that fly strips often come already printed with images of flies. In nature, a vast array of flowers take advantage of this predisposition and produce insect-like blossoms that serve as lures to draw in potential pollinators of both sexes. The dark spots on the petals of *Pelargonium tricolor,* a geranium-type shrub in South Africa, attract *Megapalpus* bee flies of both sexes, for example. Some flowers are less inclined toward equal opportunity, however. The neotropical orchid *Trichoceras parviflora,* for example, produces flowers with a remarkable resemblance to female *Paragymnomma* tachinid flies. These are attractive to male tachinids, who pollinate the flowers during their unsuccessful attempts to copulate with the flower. So the sight of a fly can have some pretty spectacular gender-specific effects in some species.

This all seems worth exploring, in both entomological and human contexts. Thinking about it, though, I'm a little frustrated that I can't actively pursue investigations of the effects of fly images on human behaviors personally. With security concerns about terrorism at new heights, it would not seem to be a propitious time for me to try to sneak into a men's bathroom at JFK Airport just to satisfy my curiosity about entomological aspects of user interface design—I doubt that my explanation for my presence there would sound plausible to any security guard. I guess, then, that men's rooms at major international terminals must remain no-fly zones for me, at least for the time being.

the unslakable mosquito

ONE INDICATION OF THE DEPTH of human animosity toward mosquitoes is the existence of the World Championship of Mosquito Killing in Pelkosenniemi, Finland. Basically, this is a competition open to all comers who are challenged to kill as many mosquitoes as possible with their bare hands within a five-minute period in an area 100 × 300 square meters. The current record of twenty-one is held by Henri Pellonpää, who in 1995 shattered the previous record of seven. All told, 370 mosquitoes bought the farm during the 1995 two-day slapfest. Although at first blush the number may appear low, particularly for the mosquito-friendly northern climes, the death toll is influenced by the fact that whenever a crowd of people assembles to cheer on the competitors their exhalations tend to draw the mosquitoes away from the main event (Cassingham 1995).

The Mosquito Killing Championship (an invention of Kai Kullervo Salmijärvi, a local businessman, in 1993) doesn't specify how the mosquitoes are to be killed—just that they must meet their fate free of insecticides and mechanical devices. It's likely the method of choice is the basic, time-honored swat. Pellonpää's record was challenged at Italy's first official mosquito-swatting competition in August, 2000, during which contestants have fifteen minutes to kill as many mosquitoes as they can. The winner of the "golden mosquito" was Christian Rizatto, who dispatched twenty-three mosquitoes.

Despite the obvious efficacy of slapping, there are much more creative ways to kill mosquitoes, if you believe Internet sites such as A Medical Professional Guide to Fascinating Mosquito Facts. According to this medical professional, Josh Stone, "One way to kill a mosquito, if you happen to catch it biting you on a convenient location such as the bicep of the arm, is to tense your skin to trap its little proboscis in your skin, then flex your bicep muscle. This apparently causes the mosquito to burst because of the pressure from your blood vessel, kind of like if you tried to drink from a fire hose."

This story is widely distributed and even appears in otherwise authoritative sources, including an article in *Discover* magazine from August, 1997, titled "Why Mosquitoes Suck." Because this article couched the description of this mode of execution in evasive language (with many qualifying words such as "maybe" and "supposedly"), Cecil Adams at The Straight Dope, a Web site noted for exploding urban legends, tackled the exploding mosquito question. He reached the conclusion initially that it is possible (August 22, 1997) but subsequently disavowed that conclusion (August 18, 2000).

I think this bit of popular wisdom persists because, for most people who have been on the wrong end of a mosquito proboscis, the image of an exploding mosquito is so very satisfying. In fact, it was one of the very first animated images of an insect to appear on a movie screen. Winsor McCay's 1912 film *How a Mosquito Operates,* one of the first line-drawing animated films ever made, depicts a dapper mosquito with top hat and briefcase who enters the room of a sleeping man to drink his fill, despite the futile efforts of the man to fend him off. Eventually, filled to capacity (SPOILER ALERT), he explodes. The image of the exploding mosquito has legs, as it were; a more recent manifestation, aired on Superbowl Sunday, 1999, was an advertisement for Tabasco sauce that depicts a mosquito sucking the blood of a man eating a Tabasco-laden slice of pizza. The mosquito flies off and ultimately explodes in a burst of flames.

If only it were as easy as flexing a muscle or wolfing down hot pepper sauce to cause a mosquito to blow up. The general scientific consensus is that it is indeed possible to cause a mosquito to explode but doing so requires severing its ventral nerve cord (Gwadz 1969). The ventral nerve cord transmits information of satiety to the mosquito's brain; when the cord is severed, the mosquito has no sense of consuming its fill. It continues to suck until it quadruples its body weight, whereupon it explodes. Moreover, even after the abdomen bursts, the mosquito continues to suck blood, which spills freely out of what remains of the back end.

Even though severing the ventral nerve cord is a sure-fire way to make a mosquito explode, it's unlikely to catch on, inasmuch as it's a little laborious to exact such a small measure of vengeance. As it turns out, even swatting a mosquito to dispatch it might be a Pyrrhic victory. A 2004 study published in the *New England Journal of Medicine* reported the case of a 57-year-old woman who died of an infection with a microsporidial parasite named *Brachiola algerae,* which normally infects only mosquitoes (Coyle et al. 2004). The unfortunate woman, who was taking a course of immunosuppressive drugs at the time to treat her rheumatoid arthritis, apparently acquired the infection as a consequence of slapping the mosquito against her skin, allowing the pathogen to gain entry into her system through the bite wound. To reduce the risk of acquiring a potentially lethal infection with this mosquito pathogen, the authors of this study accordingly recommend flicking mosquitoes rather than swatting them—although many entomologists argue that flicking allows mosquitoes to live to bite another day.

So there's no good way to kill a mosquito. The Buddhist solution, escorting any mosquitoes that enter one's home back outside (Landaw and Bodian 2003), isn't likely to catch on any faster than severing ventral nerve cords. Maybe the best bet is to call in the professionals, experts with the special weapons and tactics to deal with dangerous situations; after all, that's why they call them "SWAT" teams.

the venomous daddy longlegs

WHEN I WAS INVITED to give a plenary lecture at the International Congress of Entomology in Brisbane, Australia in August, 2004, I knew I would have to take my family along. For years, my husband and daughter have good-naturedly vacationed with me while I attended meetings in far less enticing spots; although we all had a great time in both Alpine, Texas, and West Lafayette, Indiana, for example, neither city routinely ends up on lists of the ten most popular tourist destinations. Thus, it seemed only fair, when the opportunity presented itself, to bring them to a place that large numbers of people know they want to visit, even if there's no conference in town. The problem I faced, though, was how to spend time with my family seeing what we could of Australia in the five days we were going to be there while at the same time attending to my various responsibilities during the conference. The ideal solution appeared to be to hire a private tour guide who could take us wherever we wanted to go, tell us interesting facts about the natural and cultural history of the area, and drive a car on the left side of the road without accidentally swerving into oncoming traffic or inadvertently shifting into reverse instead of signaling for a turn.

So that's how we came to know Terry of SeeMore Scenic Tours, our personal guide to Brisbane. Terry met us at our hotel the day we arrived and took us to Lone Pine Sanctuary, a nature preserve just outside the city, where tourists could get photographed holding a koala, the consummate Brisbane tourist activity. After pointing out a few highlights on the way out of town, Terry asked what brought us to Australia. When I told him I was an entomologist in Brisbane for the International Congress, he proudly informed me that the world's most venomous spider lived in Australia. "Sydney funnelweb spider, maybe, or redback spider?" I asked. "No," he replied, "it's the daddy longlegs—its venom is the deadliest in the world but its fangs are too weak to pierce the skin."

There followed an excruciating silence. What bothered me more than the technicality that daddy longlegs aren't actually spiders (they belong to the order Opiliones, not the order Araneida) was the fact that I had heard these very words many times before (albeit never before with an Australian accent). Back in the United States, "the deadly daddy longlegs" is one of the most persistent of urban legends, entirely baseless, of course. Although many do possess evil-smelling so-called repugnatorial glands, all known opilionids lack venom glands. I hesitated to say anything to Terry. For all I knew, "daddy longlegs" could be an Australian common name for something other than an opilionid—maybe even a deadly spider. After all, I hadn't been at the conference more than an hour before I discovered that an "iced coffee" in Australia (which I ordered to fend off massive jet lag) contains not ice but rather ice cream, which was just fine with me. Later in the week, though, I was dismayed to discover that a "milk shake" in Australia has no ice cream at all—it's milk and flavoring, as the name suggests; milk shaken with ice cream is called a "thick shake," and for reasons I never could ascertain, ice cream combined with soda, a beverage that is called an "ice cream soda" in the United States, is called a "spider" in Australia, further adding to the arachnolinguistic confusion.

In the United States, there's also some confusion about what a daddy longlegs is. Species in the family Pholcidae, which are true spiders, are sometimes referred to as "daddy-longlegs spiders." In Britain, flies in the family Tipulidae, known as crane flies in the United States, are often called "daddy longlegs" as well. While I was mulling over all of this, Terry noticed the prolonged silence and asked if something was wrong. I mumbled something about American daddy longlegs and then, in the hope of moving the conversation into an area about which I truly knew nothing at all, asked him to explain the rules of Australian football.

Mercifully, we didn't see any daddy longlegs of any kind on any of the subsequent trips we took with Terry, although we did see one dead funnelweb spider in a jar in a restaurant and some silk-spinning glowworms in the genus *Arachnocampa* outside Lamington National Forest. When I returned home I checked the literature on and off the Internet for whatever I could find about Australian poisonous daddy longlegs. I found that I was hardly the first entomologist to wonder about deadly daddy long-legs. In a letter published in *Natural History*, Rogelio Macias-

Ordonez (2001) responded to the query "I've been told that daddy longlegs are poisonous but have mouthparts too tiny to inflict wounds in humans. Is this true?" with the speculation that this urban legend probably arose when "at some point, an article on a group of somewhat poisonous Australian spiders that are also called daddy longlegs was picked up by the U.S. media and the creature was interpreted to be our own harvestman." Richard S. Vetter and P. Kirk Visscher (2004) were more emphatic in definitively debunking the myth, even in Australia: "This tale has been lurking around for years. I have heard it repeatedly in the United States, and even heard a schoolteacher misinforming her class at a museum in Brisbane, Australia." And finally, I found the ultimately authoritative site—the Australian Museum itself addressed the issue on its Spider FAQ Web site (as it were):

> There is no evidence in the scientific literature to suggest that Daddy-long-legs spiders are dangerously venomous . . . The jaw bases are fused together, giving the fangs a narrow gape that would make attempts to bite through human skin ineffective. However, Daddy-long-legs Spiders can kill and eat other spiders, including Redback Spiders whose venom can be fatal to humans. Perhaps this is the origin of the rumour that Daddy-long-legs are the most venomous spiders in the world.

Although I failed to turn up deadly daddy longleg venom in my Internet search, I stumbled across another distinctive feature of daddy longlegs that had escaped my notice up to that point in time. Evidently, although his mouthparts are popularly thought to be tiny, the male daddy longlegs in reality is much more impressively endowed in the genitalia department. In a paper published in the journal *Nature* titled "Preserved organs of Devonian harvestmen," Dunlop and colleagues (2003) reported finding a

400 million-year-old fossil harvestman in the Rhynie chert fossil deposits of Scotland that is clearly equipped with a male intromittent organ, or penis. This organ, together with the tracheae, or breathing tubes, and the ovipositor, or egg-laying equipment found in female specimens, is noteworthy because it provides evidence of a terrestrial existence. Intromittent organs aren't a necessity for aquatic organisms, which can discharge their sperm into an aqueous media without fear of desiccation and viability loss.

The popular press found the organ noteworthy as well, but apparently for different reasons. On the NationalGeographic.com website, John Pickrell (2003) reported the discovery of "'probably the oldest' penis found" in a spider fossil. The article itself correctly identified the fossil as a "harvestmen [sic], a non-web-spinning arachnid" and not a spider, but the focus of the article wasn't really on the challenge of colonizing terrestrial environments in the Devonian era. Although the article mentioned the tracheae, or respiratory structures, as well, these anatomical attributes were clearly of secondary status; the main emphasis of the story was indisputably the fact that this was not only a very old penis but was possibly the world's very first penis. Perhaps anticipating that their audience was more interested in sex than breathing, the story on the *National Geographic* Web site concludes with a quotation from Paul Selden, president of the International Society of Arachnology: "These type of harvestmen 'have relatively large genitalia, compared to their body size,' said Selden—the fossil male has a penis two-thirds the length of his body. 'I suppose it is to get past those long legs,' said Selden."

It amazes me that this story isn't the one circulating among the public, although perhaps some subliminal recognition of this feature of their anatomy is the reason that opilionids are universally known in the English-speaking world by male epithets, such as "daddy longlegs," "harvestmen," or "grandfather grey-beards." But the public might never have had the opportunity to acknowledge the preeminence of daddy longlegs fossil equipment; a mere month after the *Nature* article appeared, a paper by Siveter and colleagues, somewhat unexpectedly titled "An ostracode crustacean with soft parts from the Lower Silurian," was published in *Science* describing what BBC Online called the "oldest male fossil animal yet discovered"—a 425-million-year-old ostracod, a tiny crustacean sometimes called a seed shrimp. This ancient creature was sufficiently well-endowed as to inspire the name *Colymbosathon ecplecticos,* which translates to mean "amazing swimmer with a large penis." Nicholas Wade of the *New York Times* heralded the finding with the headline, "The archaeology of maleness reaches back . . . and back again" and many news Web sites, including BBC News Online, brought the update to an eager public anxious to keep abreast of late-breaking news related to preserved organs.

Frankly, I don't know why ancient intromittent organs merit so many headlines in such high-profile venues. But, then again, I don't have what *all* of the authors of *all* of these articles in *Nature, Science,* the *New York Times,* the *BBC News* Web site, the *National Geographic* Web site, and many other online news sites that carried these stories have. That's right—I don't have a Y chromosome, so I guess I'll never understand what all of the fuss is about.

the wing-flapping chaos butterfly

IN AUGUST 2003, a massive power failure plunged much of the eastern United States into darkness, disrupting traffic, emergency services, food preparation, medical-care delivery, and life in general for millions of people. A crisis of such enormous proportions hardly seemed the time to think about butterflies, yet the August 15, 2003 front-page story about the crisis in the *San Francisco Chronicle* was entitled, "How a butterfly's wing can bring down Goliath." Keay Davidson, a science writer for the paper, was of course referring to chaos theory—the idea that, in a complex system, such as an overloaded and antiquated power grid, an infinitesimal change can bring about a total collapse of the system. As for the butterfly effect itself, Davidson explained, "In the 1960s MIT meteorologist Edward Lorenz popularized the notion of the butterfly effect. An infinitesimal shift in the weather —say, the turbulence caused by a butterfly flapping its wing—can set in motion atmospheric events that climax in a hurricane. Such events are for all practical purposes unpredictable."

As a lepidopterist of sorts, I was intrigued by the butterfly reference and by Professor Lorenz, so I thought the metaphor warranted further investigation. As it turns out, Lorenz experimented with computer simulations of weather on, entomologically enough, an early computer called a "Royal McBee." He devised a series of twelve differential equations to account for various meteorological phenomena; for his model, he entered a series of variables and then ran recursive equations to generate different outcomes. One eventful day, in an effort to recreate a particular weather pattern, Lorenz entered the values recorded on a printout from the middle of the earlier run, but once the next run had completed his course, obtained a different outcome. The difference was due to the fact that the program calculated values to six significant digits, but the printout displayed values with only three significant digits. Although the difference between the two runs was tiny (one part in one thousand due to rounding error), because of the iterative nature of the calculations, the tiny error had been amplified until the ultimate outcome was completely different.

Lorenz recognized that this "sensitive dependence on initial conditions" might have broader implications. He presented the concept rather obliquely in a paper delivered in 1963 to the New York Academy of Sciences, in which he quoted a fellow meteorologist as remarking, "If the theory were correct, one flap of a seagull's wings would be enough to alter the course of the weather forever." By December 1972, in a talk presented at the American Association for the Advancement of Science (AAAS) in Washington, DC, the seagull had become a butterfly and the concept moved front and center to the title of the paper— "Predictability: Does the flap of a butterfly's wings in Brazil set off a tornado in Texas?" And thus an entomological metaphor was born.

This does not imply that an entomological metaphor persists unchanged. The general public is for the most part blissfully unaware of the proceedings of AAAS meetings—Lorenz's phrase and underlying concept went mainstream as a consequence of the publication of the best-selling popular science book *Chaos: Making a New Science,* by James Gleick, a writer for the *New York Times.* As is the case with many best-selling popular science books, the problem is that more people seem to have bought the book than actually read the book. A review of another book on chaos theory, *Turbulent Mirror* by F. David Peat, quotes the "now-famous chaos aphorism that the flutter of a butterfly's wing in Hong Kong can change the weather in New York." On a Web page dedicated to explaining computational physics, the "aphorism" was described as a "cliché" and quoted as "the flutter of a butterfly wing in Lima, Peru can affect the weather in Toronto a month later." According to the Web site of Wolfram, a mathematical software company, "a butterfly flapping its wings in Tahiti can, in theory, produce a tornado in Kansas." On a Web site about lithic technology (archeological stone tools), the butterfly effect is described as "the parable of the flapping of a butterfly's wings that creates a minor air current in China, that adds to the accumulative effect in global wind systems, that ends with a hurricane in the Caribbean." Yet another site, butterflyeffect.org, puts the butterfly in Europe: "A butterfly flapping its wings in London can, in principle, cause a subsequent hurricane in the

Philippines." According to a site maintained by the Johns Hopkins University Department of Physics and Astronomy, the ultimate effects are quite localized: "a butterfly flapping its wings in South America can affect the weather in Central Park." In an article on portfolio optimization on a Web site called CiteSeer, David Nawrocki defines the butterfly effect as "the flapping of a butterfly's wings in Beijing [that] will work its way through the system and result in a tornado in Oklahoma." And Dudley Smith, president and CEO of the World Association of Management Consulting Firms, addressed the 1996 world conference of the association in Yokohama, Japan with, "We are no better at guessing tomorrow's weather than we are at foretelling the millenium . . . A butterfly in Java waves its wings and, as a result, the weather in Chicago turns nasty."

Although the specifics vary, there's a general metaphorical pattern—butterflies flap their wings in exotic locales and bad weather results in more mundane places (as often as not, it seems, in the Midwest for some reason). Whether an actual butterfly wing could generate sufficient force to effect meteorological change isn't really the issue. As far as I can tell, such forces are rarely actually measured. I couldn't find any estimates of the force of a butterfly's wingbeat in the entomological literature; Wilkin and Williams (1993) did estimate the instantaneous vertical and horizontal forces on a moth, however. A sphingid hawkmoth weighing 14.7 millinewtons flying with a wind moving at 3.36 meters per second generates a downstroke peak of 70 millinewtons, and aerodynamic power output between 21.6 and 30.0 Watts per kilogram body mass. A millinewton is one-thousandth the force needed to accelerate a mass of one kilogram (about 2.2 pounds) by one meter per second per second. How this relates to the weather in Peoria I couldn't tell you.

But it may not matter just exactly how much force a butterfly downstroke generates. Now, even the meteorological implications of the butterfly effect have been called into question. According to chaos theory, small errors are magnified into massive effects over time; but, according to mathematician David Orrell at the University College of London, with weather forecasts initial small errors should follow the square-root law—they should become large very quickly and then slow down, so that accurate weather prediction is at least theoretically possible for short-term forecasts.

Notwithstanding, the butterfly effect is here to stay—it's mentioned in the lyrics of the song "Butterfly Wings" by the mid-1990s industrial rock group Machines of Loving Grace, it's the name of an Australian alternative rock band, it's even the title of a movie about time travel. If there's any pop-culture evidence for the unpredictability of events, it's the casting in this movie of Ashton Kutcher, of MTV's *Punk'd* and the cult hit, *Dude, Where's My Car?* in a dramatic lead. As for Lorenz, he long ago moved away from weather prediction onto other mathematical challenges; among his contributions is the so-called Lorenz attractor, from the realm of fluid dynamics. Using Navier-Stokes equations and such variables as Prandtl numbers (ratio of the fluid viscosity to thermal conductivity), temperature, and physical dimensions of the container holding the gaseous system, Lorenz devised a series of differential equations designed to predict the motion of gases enclosed and heated in a box. When plotted, the differential equations generate a structure called the Lorenz attractor:

> Instead of a simple geometric structure or even a complex curve, the structure now known as the Lorenz Attractor weaves in and out of itself. Projected on the X-Z plane, the attractor looks like a butterfly; on the Y-Z plane, it resembles an owl mask. The X-Y projection is useful mainly for glimpsing

the three-dimensionality of the attractor; it looks something like two paper plates, on parallel but different planes, connected by a strand of string. As the Lorenz Attractor is plotted, a strand will be drawn from one point, and will start weaving the outline of the right butterfly wing. Then it swirls over to the left wing and draws its center. The attractor will continue weaving back and forth between the two wings, its motion seemingly random, its very action mirroring the chaos which drives the process. (Ho 1995)

You can go to the Caltech Web site and see the attractor plotted. I don't know—it doesn't look like any species of butterfly I've ever seen. It does look like an owl mask to me, though. Maybe even a seagull, if you look at it the right way.

the x-ray-induced giant insect

MOVIE BIOLOGY OFTEN RUNS at variance with real-life biology; such is invariably the case with mutations. In the movies, mutations in insects, whether they're induced by atomic radiation (1950s), toxic waste (1970s), or genetic engineering (1990s), seemingly invariably lead to gigantism. Entomologists have a hard time taking such films seriously, inasmuch as there are several sound biological reasons we're in no immediate danger of attack by giant cockroaches with six-foot wingspans, despite what you might think when you turn on the lights in your kitchen. Among other things, insects don't breathe the same way we do. They have tracheae—holes in the sides of their body—that lead into a complex set of tubes and ducts that deliver oxygen to all parts of the body. A six-foot insect would require so much internal ductwork there would be little room inside for such vital organs as guts or hearts or brains. Another problem has to do with the fact that insects molt—shed their skins to increase in size—and it takes a while for their tough external skeleton to harden. To a small insect, gravity is negligible—but to a six-foot insect, the pull of gravity would be so strong that it would collapse in on itself before its external skeleton had time to harden.

In addition, the reality of mutations is that, by and large, mutants are a sorry lot. Most genetic aberrations lead not to enormous increases in size and strength but rather to substantial reductions in viability. Take, for example, *Drosophila melanogaster,* the fruit fly of thousands of high-school genetics classes. This species is so prone to mutations that it has proved to be an ideal subject for genetic studies. Since Thomas Morgan first discovered the convenience of working with an insect that feeds on rotting fruit and that reliably reproduces every ten days, fruit flies have been bombarded with X-rays, chemicals, and other forces in an effort to disrupt their DNA. At no time during nearly a century of work with fruit flies has any mutation led to a fruit fly of immense proportions. Some of the mutants, though, are pretty freaky. Proboscipedia, for example, is a mutation in which the mouthparts of the fly are replaced by legs.

Fruit fly mutations are the main reason I had problems in genetics class in college. The class should have been easy. The material was fairly straightforward and the professor was terrific. My problems stemmed from the fact that there were just too many distractions. For one thing, it didn't help that I used to sit next to a fellow biology major on whom I had a terrible crush. Despite my best efforts, he never seemed to acknowledge or even recognize my complete infatuation. I didn't discover until a year later that he was gay. This discovery, by the way, came two years after I discovered that the anthropology major on whom I had a crush a year earlier was gay, a year before I discovered that the ornithology graduate student on whom I had a crush was gay, and two years before I discovered, in graduate school, that the medieval Icelandic history graduate student on whom I had a crush was gay. When, as an assistant professor, I finally met the man I would eventually marry, I assumed, since I liked him so much, that he must be gay, and I'm embarrassed to admit four years passed before that impression was corrected.

The other thing I found exceedingly distracting during genetics class was studying mutations. Maybe it was because the course was taught by a *Drosophila* geneticist, but it seemed that fruit flies had more than their share of unfortunate genetic peculiarities. I found I couldn't listen to a lecture about these mutants or read a problem set in our otherwise dust-dry textbook without giggling. Among my favorites at the time were *Curly, plum, dumpy, shaven, interrupted, doublesex,* and *Prune-killer.* The names struck me as emanating from some sort of parallel-universe version of *Snow White and the Seven Dwarves* (with *white* of course being the first and foremost mutant, described by T. H. Morgan himself in 1910). Whenever I should have been listening to lectures or studying the textbook in preparation for exams, I found myself instead mentally concocting etymologically amusing but genetically improbable crosses: *raspberry lozenge? Bent blade?* Genetics with other organisms at the time simply couldn't compare. Our textbook's index listed fewer than two dozen *Escherichia coli* bacterial mutants and absolutely none had names that inspired creative daydreaming. The bread mold *Neurospora crassa* was only slightly better (offering *poky* and *snowflake*), and mutant mice

were a disappointment, with only one of eighteen mutations—
Danforth's short tail—possessing even the slightest hint of
whimsy.

I'm just grateful I took genetics over thirty years ago; today, I
would lose all semblance of focus after the first ten minutes of
any *Drosophila* genetics lecture. New model organisms, such as
the plant mouse-ear cress *(Arabidopsis thaliana)* and zebrafish,
have entered the literature in the past three decades with de-
corum. Mutations for the most part are neatly numbered and
coded. *Drosophila* geneticists, however, have tapped into all of human knowledge for naming inspiration. The Web site Flybase
provides a comprehensive database of all *Drosophila melanogaster*–
related genetics and molecular biology. The site has provided me
with endless hours of entertainment. At the click of a mouse
(not of the *Danforth's short tail* variety), any interested party can
discover the etymology of over 400 gene names. For those unwilling to devote hours of displacement activity to unearthing
etymologies, there's also the site Flynome, which recounts the
origins of a select group of interesting *Drosophila* gene names.

Learning about *D. melanogaster* mutants is a four-year liberal
arts education crammed into a single genome. It's not surprising
that fly geneticists are conversant with biology and have named
mutations for resemblances to animals both extant and extinct,
including *pangolin, hedgehog, armadillo, baboon, rhino,* and, for
wing mutants, *moa* and *piopio* (which are, or were, in the case
of the extinct moa, wingless birds). But it is surprising, I guess,
that there are mutants with names that require a knowledge of
European history; *tudor, staufen, vasa,* and *valois,* for example, are
all lethal "grandchildless" mutants named for European royal
families that ended without issue. Cell-division mutations in
which nuclei or parts thereof fail to reach the posterior pole of
the cell are named *barentsz, scott of the Arctic,* and *shackleford* in

memory of those explorers who also failed to reach a Pole (albeit a global and not a cellular one). Astronomical science is represented—*hale bopp* and *Schumacher-Levy*, named for twentieth-century comets, are two mutations producing developmental comet-shaped abnormalities in elongating spermatids. Astronomical science fiction is even represented. Plot details in various episodes of *Star Trek* inspired the naming of *klingon* and *tribbles*. Mutants invoking the great canon of Western literature (*prospero*, *hamlet*, *malvolio*, and *capulet* from Shakespeare's opus) share a genome if not a chromosome with fictional cultural icons.

Television has probably provided about as much metaphorical fodder as has the entire western European literary tradition; *maggie*, for example, is a mutation that arrests larval development in first instar, much as Maggie Simpson has remained an infant for nineteen seasons of the animated series *The Simpsons*. Mutant *kenny* flies with immune-system defects are prone to early demise, much as is Kenny on *South Park*, who reliably dies before the end of each episode. Movies, too, creep into the genome; *indy*, a mutation that extends lifespan beyond the norm, is actually an acronym for "I'm Not Dead Yet," a line from *Monty Python and the Holy Grail* spoken by an ostensibly dead plague victim being carted away prematurely for burial.

Even food can be fodder for *Drosophila* geneticists—the list of mutations that cause defects in oogenesis, or egg formation, for example, reads like instructions for a short-order cook, with *fried*, *omelet*, *sunnyside up*, *hard boiled*, *soft boiled*, *poached*, and *benedict* (inventorying, as Morris et al. 2003 describe, "the unfortunate fates commonly met by eggs"). Beyond eggs, other breakfast items inspiring *Drosophila* geneticists include *currant bun*, *clootie dumpling*, and *spotted dick*.

Clearly, naming genes is an international effort and there's nothing better for fostering an appreciation of other cultures than learning a little about their language. Genes have acquired names in Hebrew *(keren)*, Catalan *(capicua)*, Yiddish *(nebbish)*, Chinese *(hu li fai shao)*, Russian *(zlodny)*, and French *(tout-velu)*, among other languages. Thanks in large part to the enormous influence of Nobel laureates Eric Wieschaus and Christine Nusslein-Vollhard, parts of the *Drosophila* genome look like an introductory German Vokabelnprüfung (with, for example, *hitzschlag, kastchen, klarsicht, klumpfuss, klotzchen, kelch, krotzkopf verkehrt, verkerht, mochtegern, toll, zerknullt,* and *weniger*). Beyond mere words are arcane cultural references. The mating behavior mutant *la voile et la vapeur,* in which male heterozygotes court flies of both sexes, owes its name to a French slang expression that's roughly the equivalent of "AC-DC." The German *bruchpilot,* which means "crash pilot," describes a mutant that survives despite impaired flight capacity and invokes a 1941 German cult-film favorite, *Quax, der Bruchpilot.* There's even a mutation named in an extinct language, Nahuatl; *matopopetl* means "balls" and, according to Flybase, "apparently refers to the many balls of cells found in 'topi' mutant testes (Perezgasga et al. 2004)"; I have a sneaking suspicion that it's actually a bilingual pun.

Drosophila geneticists have prevailed in their learned naming practices despite objections from other, more staid geneticists. Under pressure, some names in questionable taste have been revised. In 1963, the mutation that causes male flies to court other males was named *fruity,* but political correctness led a name change to the equally apt but less offensive *fruitless* years later (Broadfoot 2001). The propriety of naming learning-defect mutants after vegetables (e.g., *turnip, radish, rutabaga*) drew criticism in more politically correct decades. And at least one mutant name, *kuzbanian,* named in reference to the Koozbanian alien puppet creatures (equipped with supernumerary bristles) on *The Muppet Show,* almost elicited a lawsuit for copyright infringement until the spelling was changed.

In part to avoid such problems, but mostly to eliminate redundancies as homologues are identified and functions are clarified, there are now efforts afoot to standardize gene nomenclature across all organisms. The stated goal of the Gene Ontology Consortium is to "produce a dynamic, controlled vocabulary that can be applied to all eukaryotes even as knowledge of gene and protein roles in cells is accumulating and changing" (Ashburner et al. 2000). I hope, though, that *Drosophila* geneticists stay true to their tradition. They're real Renaissance scholars, living the spirit of multidisciplinarity. To maintain such dazzling breadth of knowledge in the face of social and scientific pressures to conform is a challenge these days that takes certain anatomical attributes—*matopopetl,* if you will, a trait notably lacking in the aptly named fly mutant *ken and barbie.*

the yogurt beetle

THE INTERNET IS A NOTORIOUSLY unreliable source of information, primarily because of its openness and accessibility. Fortunately, it is, to some extent, self-correcting. That's the idea behind Wikipedia, the open-source encyclopedia that is effectively proofread by millions of potential editors. Statistically speaking, by sheer chance a legitimate authority on any given subject will at some point likely encounter some inaccurate information and, if he or she is technically savvy enough, will correct it. In addition, there are sites dedicated to serving as authoritative sources on all kinds of information. But, sad to say, some of these sites are the sources and perpetuators of misinformation. That was the situation for a number of years in the case of yogurt beetles.

Snopes.com is a Web site that bills itself as an authoritative source on the substance of urban myths and legends. Occasionally, they'll tackle insect subjects and usually do, in my opinion as an entomologist, a credible job. One glaring exception was their entry titled "Red Red Whine." When I first came across this site several years ago, I found this text:

> Claim: The food colorants cochineal and carmine are made from ground beetles.
> Status: True.
> Example: Collected via e-mail, 2001.
> There is a book out very recently that claims the red color of strawberry milkshakes comes from a tropical beetle ground up for its red coloring.

Snopes.com did in fact confirm that cochineal and carmine are "derived from the crushed carcasses of a particular South and Central American beetle," specifically "from the female *Dactylopius coccus,* a beetle that inhabits a type of cactus known as *Opuntia."*

Cochineal is indeed a pigment produced by *Dactylopius coccus,* which feeds on species of cacti in the genus *Opuntia.* It is also true that the Aztec Indians used these insects as a source of pigment, as Snopes.com reported, and that the Spanish conquistadors, recognizing the value of the colorant, took them back to Europe, where cochineal quickly became a valuable article of commerce. It's true that it is used in a variety of products to produce a red color, including strawberry milkshakes as well as yogurt. According to the site, then, reddish dairy products may contain ground-up beetles along with calcium and protein. Not everyone is a fan. Some people are allergic, and Orthodox Jews consider products colored with cochineal as unkosher. Score many points for Snopes.com's reporting thus far, but subtract some heavy-duty points on the very first count; cochineal is most definitely and emphatically NOT from a beetle. *Dactylopius coccus* is in fact a member of the order Hemiptera, suborder Homoptera—it's a scale insect and is no closer to a beetle than a squirrel is to a bat. The authoritative sources on the information presented at the site were articles from three newspapers—*The Montreal Gazette, the Denver Post,* and the *Bergen County Record.*

It didn't bother me that the Snopes.com site was perpetuating an entomological error—after it, it's a great source of accurate information about other misperceptions—but what did bother me is the fact that, once their error was pointed out to them, they were extraordinarily reluctant to acknowledge it. I wasn't the one who called them on it. I have an aversion to communicating with strangers via the Internet, likely a vestige of parental cautions against talking to strangers. It was a fellow entomologist named Steve Bambara. Steve is an extension entomologist in North Carolina, which means he answers questions about insects from the public for a living. I don't know how he first encountered "Red Red Whine," but after he did he sent an email message to Snopes.com with the correction. He received in response a note that, according to a dictionary such as *Webster's,* it's perfectly appropriate to call anything that resembles a beetle a beetle.

I've had many discussions with newspaper editors and trade journal editors about the fact that no standard dictionary is an authoritative source for scientific terms. The Entomological Society of America, for example, went to a lot of trouble to assemble a list of official common names for insects so that people wouldn't have to wrestle with Latin but could still be precise about what they're talking about. This isn't just pedantry—with over a million species to consider, entomologists need to be precise in talking about them, whether in Latin or in English.

Although *Webster's* can be an authoritative source for resolving Scrabble controversies, it's less authoritative on scientific matters. On the face of it, the second definition of "beetle" is ridiculous. Among other things, it's a major stretch to say that cochineal scale insects resemble beetles. Beetles are defined by biting mouthparts and by a hardened pair of front wings, called elytra, whereas cochineal scale insects have sucking mouthparts and only the male has wings at all, which are completely membranous. Sure, they resemble beetles in having six legs, although adult females don't have any legs, but so do about a million other species. For that matter, there are over 350,000 species of beetles, so there is arguably no distinctive beetle gestalt. There are plenty of beetles that don't look particularly beetlelike. Beaver beetles, for example, are parasites that live in the fur behind the ears of beavers and resemble lice more than beetles. Staphylinids, or rove beetles, scavenge around in ant nests and look like ants. There's even a group of chrysomelid leaf beetles that look like caterpillar droppings.

Bats look more like rats than cochineal scale insects look like beetles, antlike or otherwise, but I seriously doubt that the folks at Snopes.com would be complacent about calling bats rats. To carry their argument to its logical extreme, *Webster's* defines "dog" in definition 1b as "a male of any carnivorous mammal." So by that logic, it should be perfectly legitimate to call a full-maned lion on the African veldt a dog.

Since Steve Bambara took issue with Snopes.com, at least two other entomologists told me they contacted the Web site with much the same outcome. What's puzzling about all of this is that it appears that the Snopes.com editors routinely consult experts, even entomologists. They quote authorities on the validity of breast-infesting maggots, termites in mulch from post-Katrina New Orleans, and toxic stick insects spraying acrid solutions at dogs in Texas. Well, at least I assume they were spraying at dogs; maybe male lions are stalking the streets of Waco, given the imprecision of the dictionary definition. Notwithstanding, the site has been updated and the claim now reads, "The food colorants cochineal and carmine are made from ground bugs," and the word "beetle" no longer appears anywhere in the entry aside from the initial strawberry milkshake example. Score one for the entomologists.

Snopes.com concludes the entry with the statement, "Western society eschews (rather than chews) bugs, hence the widespread 'Ewww!' reaction to the news that some of our favorite foods contain insect extract." I wonder whether Snopes.com is aware that, in addition to carmine/cochineal, another product of a homopteran often graces our plate. It is a food ingredient euphemistically called "resinous glaze," "confectioner's glaze," or "pharmaceutical glaze"—the shiny coating on candies, pills, tablets, and capsules that enhances the appearance of the product, improves its shelf life, and keeps out moisture. That's actually shellac, derived from the resinous secretions of *Kerria lacca,* the lac bug. These bugs form enormous aggregations on their fig and banyan host trees in India. The aggregated mass of resin is scraped off and chemically processed to produce shellac, which, in addition to candy and medicine, is used to polish furniture, bowling alleys, violins, and dental equipment. I was thinking of sending a note to that effect to Snopes.com, but I really don't want to open that can of worms (as it were).

the zapper bug

IT'S AN ODD REFLECTION of American values and priorities that among the very first uses to which electricity was put was to devise a means of killing people. A dentist from Buffalo, New York is credited with being the first to recognize the potential of, well, potential for executing people (Brandon 1999). In 1881, Alfred Porter Southwick had read accounts in the local newspaper of the fate of the unfortunate George L. Smith, a dockworker who expired quickly after curiosity and possibly excessive alcohol consumption led him to place both of his hands on a live 4,800 pound electric generator. An amateur inventor,

Southwick toyed with the idea of exploiting electricity as a dental anesthetic but ultimately conceived the notion of using electricity as an alternative to hanging for capital punishment. As a dentist, he was accustomed to having his patients sit in chairs, so in retrospect, it's not at all surprising that he suggested delivering a fatal shock to a person sitting in a chair for execution.

A functional chair was first developed by Harold P. Brown, a disciple of Thomas Edison, after he publicly electrocuted numerous small animals with alternating current to demonstrate its lethality. As a result of enthusiastic advocacy by the governor of New York, David B. Hill, as well as the state's Electrical Death Commission (on which Southwick served as a member), electrocution for capital punishment was legalized on January 1, 1889. William Kemmler, of Auburn, New York, who had been sentenced to death for the hatchet murder of his wife, earned lasting if dubious fame on August 6, 1890 as the first criminal to be electrocuted for the commission of a crime. The event didn't quite go off as planned—the first jolt, 17 seconds of 1,000 volts, failed to kill Kemmler. The second jolt, of about 1,700 volts, killed him only after setting him ablaze. Despite its less-than-perfect debut, the electric chair continued to be used as a putatively humane alternative to other forms of execution.

It's perhaps an even odder reflection of priorities that almost a half-century of human electrocutions accrued before anyone thought of using electricity to dispatch insects. It's not clear who coined the term "zapper," but its etymology reflects the onomatopoetic death of any small creature who ends up on the grid. The word dates back at least to 1929 as a universal sound effect for electrical shock—in Dutch, for example, it's *zappen* and in French *zapper,* or *faire du zapping.* The first insect electrocution device was invented about this time. The prototypical bug zapper might well have been the flytrap proposed by Charles G. Seefluth and John Bebiolka of Pontiac, Michigan in 1924 (U.S. Patent 1486307). They described their invention as

> an electric fly and insect destroyer or exterminator in the form of a trap of simple and novel construction and more particularly to an electrically energized contact bar exterminator, the object of which is to provide a trap for destroying flies and other obnoxious insects which when the insect comes into contact with the active elements thereof will electrocute them and thus exterminate the same in a sanitary and efficient manner, thereby obviating the necessity of employing catching means which mutilate or similarly injure the flies, or of providing sticky fly paper or poisonous materials which are unsanitary and dangerous to use.

An essential component of the design was an incandescent light bulb equipped with reflectors to "aid in luring insects to the trap . . . in the nighttime as well as in the day time." Another innovation was the arrangement of the active elements in an insulated frame so as to reduce the risk of shock to human handlers.

It's interesting to note that one of the principal selling points of the new insect electrocution device—that it could kill flies without mutilating or otherwise injuring them—was not unlike the argument advanced for substituting electrocution for hanging for the purpose of capital punishment. The essence of the bug zapper has remained unchanged for the past eight decades: the electrically charged wire grid provides a charge of about 1,000 volts when an insect comes close enough to complete a circuit (within 1 millimeter of a grid wire, because an arc can form in an air gap).

This isn't to say that there haven't been refinements. At least 200 patents have been filed since 1924 to improve the process of insect electrocution. In 1934, for example, William F. Folmer and Harrison L. Chapin added baffles to the design (Patent Number 1,962,439) to capitalize on the spiraling flight of insects attracted to light:

> It is a fact of common observation that moths and other night flying insects in their erratic flight about a lamp take a generally circular course of narrowing diameter until they finally reach the light center, though there seems to be so little preconception about their intention that they will often veer off into the shadows and disappear for no apparent reason at all. There is a theory among entomologists that their reactions to light are subconscious or purely mechanical in that light affects their nerves and muscles automatically rather than breeding in them a desire to reach the light through ocular perception. However this may be, I provide the exterminator with means which extends the scope of its effectiveness to intercept revolving insect bodies that might otherwise escape.

Other practical improvements included components to attach electrocution devices to tractors or other wheeled vehicles (3846932 Bialobrzeski 43/138) and changed the orientation from vertical to horizontal (Iannini 3894351). To keep up with the times, bug zappers have gone high-tech. United States Patent D522085 5343652, for example, is for an "apparatus for laser pest control, which, . . . uses a laser beam which is scanned over a defined area and incapacitates sensory organs of various pests when they enter the defined area. Such a pest control system uses a laser source in cooperation with a scanner which then repetitively scans the laser beam throughout the defined area. Any pest which wanders into this area, or is attracted into this area, is likely to sense the laser beam, typically through its eyes or light spot. The laser beam is of sufficient energy to destroy the sensory organ and incapacitate the pest." And they've even gone organic—the most recent patent filed for an insect extermination device is from March 20, 2008, for an "organic insect extermination lamp" (United States Patent 20080066372) "which . . . comprises a fixture having a power supply and a light source for attracting mosquitoes and other biting insects and at least one container for holding a natural exterminating substance, wherein the natural exterminating substance evaporates natural exterminating vapors for killing the mosquitoes and other biting insects. The natural exterminating substance may comprise any allyl sulfide emulsion or any other organic compound. An allyl sulfide emulsion may contain garlic oil, garlic paste, garlic emulsion, crushed fresh garlic, or other forms of natural killing compounds." Interestingly, garlic has never been explored as a substitute for electricity in human executions.

Some updates have been less practical—in the Green Talking Bug Zapper, electrocution sets off amusing recorded phrases. Ads for the device state, "Zap annoying insects with a humorous twist! Flies and mosquitoes are greeted with one of over 15 hysterical phrases including 'That's gonna leave a mark!' and 'Goodbye, cruel world!' when they touch the zapper grid." Patent 6195932 is for a "Musical electronic insect killer—An electronic insect killer apparatus which generates a musical song, noise or display in response to detecting the electrocution of an insect." Recent advertisements indicate that homeowners can also kill insects while conserving energy. The Viatek Two in OneDigital UV Bug Zapper and LED Lantern BL01G "is rechargeable, chemical free and user can select bug zapper or LED lantern! . . . The LCD display will keep you apprised of the time, date and current temperature." And the "SolZapper Solar Bug Zapper is both a solar light and a solar bug zapper and will turn on and off automatically thanks to the light-detecting photo cell. The Solzapper is perfect for both dramatically lighting the pathway through your garden and protecting your plants."

Although bug zappers are clearly effective at dispatching insects, there's absolutely no evidence that they provide the function they are generally purchased for in suburban backyards—to kill mosquitoes and other biting flies. Multiple studies have repeatedly demonstrated that bug zappers, wherever they are deployed, fail to reduce the likelihood of being bitten by a mosquito (Nasci et al. 1983). For one thing, some mosquitoes aren't attracted to light at all; moreover, those that are attracted to a trap from a distance generally switch to host-seeking behavior when they get close and thus are rarely electrocuted. In one study conducted in Newark, Delaware in 1994, 13,789 insects were killed over a ten-week period; of these, only thirty-one (0.22 percent) were mosquitoes (Frick and Tallamy 1996). Nearly half of the insects killed were totally harmless midges and caddisflies. At the time, national bug zapper sales averaged approximately 1 million per year; assuming that every year 4 million bug zappers are deployed for even half of the summer months, bug zappers could be responsible for the deaths of 71 billion harmless or even beneficial insects. Such a death toll almost assuredly affects the structure and function of food webs in backyards, campgrounds, and recreational areas; in the words of the Green Talking Bug Zapper, that surely is "gonna leave a mark."

references

Abrahams, M. 2002. *The IgNobel Prizes: The Annals of Improbable Research*. New York: Dutton.

Abramson, C. I., S. M. Stone, R. A. Ortez, A. Luccardi, K. L. Vann, K. D. Hanig, and J. Rice. 2000. The development of an ethanol model using social insects. I. Behavior studies of the honey bee (*Apis mellifera* L.). *Alcoholism* 24: 1153–66.

Abramson, C. I., C. Sanderson, J. Painter, S. Barnett, and H. Wells. 2005. Development of an ethanol model using social insects: V. Honeybee foraging decisions under the influence of alcohol. *Alcohol* 36: 187–93.

Abramson, C. I., H. Wells, and J. Bozic. 2007. A social insect model for the study of ethanol induced behavior: The honey bee. In *Trends in alcohol abuse and alcoholism research*, ed. Rin Yoshida, 197–218. Hauppauge, NY: Nova Science Publishers.

Altschuler, D. Z., M. Crutcher, N. Culceanu, B. A. Cervantes, C. Terinte, and L. N. Sorkin. 2004. Collembola (springtails) (Arthropoda: Hexapoda: Entognatha) found in scrapings from individuals diagnosed with delusory parasitosis. *Journal of the New York Entomological Society* 112: 87–95.

Antonelli, P. J., A. Ahmadi, and A. Prevatt. 2001. Insecticidal activity of common reagents for insect foreign bodies of the ear. *Laryngoscope* 111: 15–20.

Arikawa, K., D. Suyama, and T. Fujii. 1997. Hindsight by genitalia: Photoguided copulation in butterflies. *Journal of Comparative Physiology A* 180: 295–99.

Armstrong, N. R., and J. D. Wilson. 2006. Did the "Brazilian" kill the pubic louse? *Sexually Transmitted Infections* 82(3): 265–66.

Apt, L. 1995. Flea collar anisocoria. *Archives of Ophthalmology* 113: 403.

Ashburner, M., et al. 2000. Gene ontology: Tool for the unification of biology. *Nature Genetics* 25: 25–29.

Australian National Museum. 2004. Spiders: General FAQs. http://www.amonline.net.au/spiders/resources/general.htm#venomous.

Berenbaum, M. R. 1995. *Bugs in the System: Insects and Their Impact in Human Affairs.* Reading: Addison Wesley.

———. 2002. How little we know. *NewsDay* op-ed page, Sunday, June 9. http://www.newsday.com/news/opinion/ny-vpber092738442jun09.story.

———. 2007. Losing their buzz. *New York Times,* March 2.

Bilde, T., C. Tuni, R. Elsayed, S. Pekar, and S. Toft. 2006. Death feigning in the face of sexual cannibalism. *Biology Letters* 2: 23–25.

Borror, D. J., D. M. DeLong, and C. A. Triplehorn. 1976. *An Introduction to the Study of Insects.* New York: Holt, Rinehart, and Winston.

Bozic, J., J. DiCesare, H. Wells, and C. Abramson. 2006. Ethanol levels in honeybee hemolymph resulting from alcohol ingestion. *Alcohol* 41: 281–84.

Brandon, C. 1999. *The Electric Chair: An Unnatural History.* New York: McFarland.

Bressler, K., and C. Shelton. 1993. Ear foreign-body removal: A review of 98 consecutive cases. *Laryngoscope* 103: 367–70.

Broadfoot, M. V. 2001. A gene by any other name: Whimsy and inspiration in the naming of genes. *American Scientist* 89 (November/December, no. 6): 1.

Brooke, M. W. 1881. Influence of temperature on the chirp of the cricket. *Popular Science Monthly* 20: 268.

Brunvand, J. 1981. *The Vanishing Hitchhiker: American Urban Legends and Their Meanings.* New York: W.W. Norton.

———. 1993. *The Baby Train and Other Lusty Urban Legends.* New York: W.W. Norton.

Buffon, G. 1749–1788. *Histoire naturelle, générale et particulière* (vol. 5). Paris: Imprimeries royale.

Butler, C. 1609. *The Feminine Monarchie, or, A Treatise Concerning Bees and the Due Order of Them.* Oxford: Joseph Barnes.

Carroll, R. T. 2005. Pareidolia. *The Skeptic's Dictionary.* http://skepdic
.com/pareidol.html.

Cassingham, R. 1995. *This Is True: Glow-in-the-Dark Plants Could Help Farmers.* Boulder: Freelance Communications.

Clausen, L. W. 1954. *Insect Fact and Folklore.* New York: Macmillan.

Cork, J. M. 1957. Gamma radiation and longevity of the flour beetle. *Radiation Research* 7: 551–57.

Cornwell, P. B., L. J. Crook, and J. O. Bull. 1957. Lethal and sterilizing effects of gamma radiation on insects infesting cereal commodities. *Nature* 179: 670–72.

Cowan, F. 1865. *Curious Facts in the History of Insects.* Philadelphia: J. B. Lippincott.

Coyle, C. M., L. M. Weiss, L. V. Rhodes III, A. Cali, P. M. Takvorian, D. F. Brown, G. S. Visvesvara, L. Xiao, J. Naktin, E. Young, M. Gareca, G. Colasante, and M. Wittner. 2004. Fatal myositis due to the microsporidian *Brachiola algerae,* a mosquito pathogen. *New England Journal of Medicine* 351(1): 42–47.

Culpepper, N. 1652. *The English Physitian: Or an astrologo-physical discourse of the vulgar herbs of this nation.* London: Peter Cole.

D'Angelo, J. P., and D. B. West. 2000. *Mathematical Thinking: Problem-Solving and Proofs.* 2nd ed. Upper Saddle River, NJ: Prentice-Hall.

Davey, W. P. 1919. Prolongation of life of *Tribolium confusum* apparently due to small doses of X-rays. *Journal of Experimental Zoology* 28: 447–58.

Dewaraja, R. 1987. Formicophilia, an unusual paraphilia, treated with counseling and behavior therapy. *American Journal of Psychotherapy* 41: 593–97.

Dewaraja, R., and J. Money. 1986. Transcultural sexology: formicophilia, a newly named paraphilia in a young Buddhist male. *Journal of Sex and Marital Therapy* 12: 139–45.

Dias, J. C. 1997. [Cecílio Romaña, Romaña's sign and Chagas' disease.] *Revista da Sociedade Brasileira de Medicina Tropical* 30(5): 407–13.

Dolbear, A. E. 1897. The cricket as thermometer. *American Naturalist* 31: 970–71.

Dunlop, J. A., L. I. Anderson, H. Kerp, and H. Hass. 2003. Preserved organs of Devonian harvestmen. *Nature* 425: 916.

Edes, R. T. 1898. Rate of the chirping of the tree cricket *(Oecanthus niveus)* to temperature. *American Naturalist* 33: 935–38.

Einstein, A. 1949. Why socialism? *Monthly Review* 1: 9–15.

Eisenstein, E. M. and M. J. Cohen. 1965. Learning in an isolated protho-racic insect ganglion. *Animal Behavior* 13: 104–8.

Enright, J. T. 1971. The internal clock of drunken isopods. *Zeitschrift für vergleichende Physiologie* 75: 332–46.

Fabre, J. H. 1916. *Social Life in the Insect World.* Trans. Bernard Miall. New York: Century, 1916.

Fowles, G., and G. Cassiday. 1990. *Analytical Mechanics.* 6th ed. New York: Saunders College Publishing, p. 137.

Franceschini, N. 1985. Early processing of colour and motion in a mosaic visual system. *Neuroscience Research–Supplement* 2: S17–S49.

Frick, T. B., and D. W. Tallamy. 1996. Density and diversity of nontarget insects killed by suburban electric insect traps. *Entomological News* 107: 77–82.

Frings, H., and M. Frings. 1957. The effects of temperature on chirp-rate of male cone-headed grasshoppers, *Neoconocephalus ensiger. Journal of Experimental Zoology* 134: 411–25.

Garrett, B. C. 1996. The Colorado potato beetle goes to war. Historical note no. 2. *Chemical Weapons Convention Bulletin* no. 33: 2–3.

Gemeno, C., and J. Claramunt. 2006. Sexual approach in the praying mantid *Mantis religiosa. Journal of Insect Behavior* 19: 731–40.

Gordon, D. G. 1996. *The Compleat Cockroach.* Berkeley: Ten Speed Press.

Goudey-Perriere, F., Lemonnier, F., Perriere, C., Dahmani, F. Z., and Wim-mer, Z. 2003. Is the carbamate juvenoid W-328 an insect growth regulator for the cockroach *Blaberus craniifer* Br. (Insecta, Dicty-optera)? *Pesticide Biochemistry & Physiology* 75(1–2): 47–59.

Gwadz, R. W. 1969. Regulation of blood meal size in the mosquito. *Journal of Insect Physiology* 11: 2039–44.

Haldane, J. B. S. 1928. On being the right size. In J. B. S. Haldane and J. M. Smith, eds. 1985. *On Being the Right Size and Other Essays.* New York: Oxford University Press.

Halder, G., P. Callaerts, and W. J. Gehring. 1995. Induction of ectopic eyes by targeted expression of the eyeless gene in *Drosophila. Science* 267: 1788–92.

Hanson, F. B., and F. M. Heys. 1928. The effects of radium in producing lethal mutations in *Drosophila melanogaster*. *Science* 68: 115–16.

Hardy, D. E. 1940. Studies in New World *Plecia* (Bibionidae: Diptera). Part I. *Journal of the Kansas Entomological Society* 13: 15–27.

Hase, A., 1929. Zur pathologisch-parasitologischen und epidemiologisch-hygienischen Bedeutung der Milben, insbesondere der Tyroglyphinae (Käsemilben), sowie über den sogenannten Milbenkäse. *Beiträge zur experimentellen Parasitologie* 1: 765–821.

Hetrick, L. A. 1970. Biology of the "love-bug," *Plecia nearctica* (Diptera: Bibionidae). *Florida Entomologist* 53: 23–26.

Hinton, H. E. 1973. Natural deception. Chap. 3 in *Illusion in Nature and Art*, ed. R. L. Gregory and H. Gambrich. London: Gerald Duckworth.

———. 1974. Lycaenid pupae that mimic anthropoid heads. *Journal of Entomology* (A) 49: 65–69.

Ho, Andrew. 1995. Lorenz attractor. http://www.zeuscat.com/andrew/chaos/lorenz.html.

Horridge, G. A. 1962. Learning of leg position by the ventral nerve cord in headless insects. *Proceedings of the Royal Society of London B* 157: 33–52.

Howard, L. O. 1886. The excessive voracity of the female Mantis. *Science* 8: 326.

Howard, L. O. 1899. Spider bites and kissing bugs. *Popular Science Monthly* 56: 31–42.

Johnson, J. C., T. M. Ivy, and S. K. Sakaluk. 1999. Female remating propensity contingent on sexual cannibalism in sagebrush crickets, *Cyphoderris strepitans:* A mechanism of cryptic female choice. *Behavioral Ecology* 10: 227–33.

Kamimura, Y., and Y. Matsuo. 2001. A "spare" compensates for the risk of destruction of the elongated penis of earwigs (Insecta: Dermaptera). *Naturwissenschaften* 88: 1432–1904.

Kandel, E. R., and J. H. Schwartz. 1985. *Principles of Neural Science*. 2nd ed. New York: Elsevier.

Karras, D. J., S. E. Farrell, R. A. Harrigan, F. M. Henretig, and L. Gealt. 1996. Poisoning from "Spanish fly" (cantharidin). *American Journal of Emergency Medicine* 14(5): 478–83.

Kenward, H. 1999. Pubic lice (*Pthirus pubis* L.) were present in Roman and medieval Britain. *Antiquity* 73: 911–15.

Kirby, W., and W. Spence. 1818. *Introduction to Entomology*. 4 vols. (1818–1826). London, Longman.

Kitching, I. A. 2003. Phylogeny of the death's head hawkmoths Acherontia (Laspeyres) and related genera (Lepidoptera: Sphingidae: Sphinginae: Acherontiini). *Systematic Entomology* 28: 71–88.

Koch, R. L. 2003. The multicolored Asian lady beetle, *Harmonia axyridis:* A review of its biology, uses in biological control and non-target impacts. *Journal of Insect Science* 3: 32–48.

Korte, G. 2004. Email: Democrat morphs like insect. *Cincinnati Enquirer,* May 15, 2004. http://www.enquirer.com/editions/2004/05/15/loc_cicadapolitics15.html.

Landaw, J., and S. Bodian. 2003. *Buddhism for Dummies*. New York: Wiley.

Lelito, J P. and W. O. Brown. 2006. Complicity or conflict over sexual cannibalism? Male risk taking in the praying mantis *Tenodera aridifolia sinensis. American Naturalist* 168: 263–69.

Lemon, G. W. 1783. *English etymology; or, A derivative dictionary of the English language*. London: G. Robinson.

Limbaugh, R. 2008. Queen bee Pelosi stings Hillary. *The Rush Limbaugh Show,* March 13. http://www.rushlimbaugh.com/home/daily/site_031308/content/01125111.guest.html.

Linnaeus, C. 1758. *Systema naturae per regna tria naturae* (10) 143, no. 1. Holmiae (Stockholm).

Lockhart, G. 1988. *The Weather Companion*. New York: John Wiley and Sons.

Lorenz, E. N. 1963. Deterministic nonperiodic flow. *Journal of Atmospheric Science* 20: 130–41.

Loud, L. 1990. When spiders miss their marks (work of insect trainers on the set of *Arachnophobia*). *American Film* 15: 64.

Lubbock, J. 1882. *Ants, Bees, and Wasps: A Record of Observations on the Habits of the Social Hymenoptera*. New York: D. Appleton.

Lutz, F. E. 1914. Humidity—a neglected factor in experimental work. *American Naturalist* 48S: 122–28.

Lyman, H. H. 1899. The president's annual address. Thirtieth annual report of the Entomological Society of Ontario. Toronto: Warwick Bros. & Rutter, pp. 21–30.

Macias-Ordonez, R. 2001. Daddy longlegs. *Natural History* (February): 12.

Magnan, A. 1934. *Le vol des insectes.* Paris: Hermann and Cle.

Mavin, S. 2008. Queen bees, wannabees and afraid to bees: No more 'best enemies' for women in management? *British Journal of Management* 19 (s1): S75–S84.

Moncayo, A. 2003. Chagas disease: Current epidemiological trends after the interruption of vectorial and transfusional transmission in the Southern Cone countries. *Memórias do Instituto Oswaldo Cruz* 98: 577–91.

Morris, J. Z., C. Navarro, and R. Lehmann. 2003. Identification and analysis of mutations in *bob, Doa* and eight new genes required for oocyte specification and development in *Drosophila melanogaster. Genetics* 164: 1435–46.

Muffet, T. 1634. *Theater of Insects appended to Edward Topsell History of Fourefooted Beasts and Serpents.* London: William Iaggard.

Muller, H. J. 1927. Artifical transmutation of the gene. *Science* 66: 84–87.

Nakagawa, T., and E. Eguchi. 1994. Differences in flicker fusion frequencies of the five spectral photoreceptor types in the swallowtail butterfly's compound eye. *Zoological Science* 11: 759–62.

Nasci, R. S., C. W. Harris, and C. K. Porter. 1983. Failure of an insect electrocuting device to reduce mosquito biting. *Mosquito News* 43: 180–84.

Newman, B. 1997. Apple turnover: Dutch are invading JFK Arrivals building and none too soon. *Wall Street Journal,* May 13, 1997, p. A1.

Nickolls, L. C., and D. Teare. 1954. Poisoning by cantharidin. *British Medical Journal* 2(4901): 1384–86.

Okamura, C. 1980. Period of the Far Eastern mini-creatures. Original Report of the Okamura Fossil Laboratory, no. 14.

———. 1983. New facts: Homo and all vertebrates were born simultaneously in the former Paleozoic in Japan. Original Report of the Okamura Fossil Laboratory, no. 15.

O'Toole, K., P. M. Paris, R. D. Stewart, and R. Martinez. 1985. Removing cockroaches from the auditory canal: A controlled trial. *New England Journal of Medicine* 312: 1197.

Overstreet, R. 2003. Presidential address: Flavor buds and other delights. *Journal of Parasitology* 89: 1093–1107.

Patek, S. N., J. E. Baio, B. L. Fisher, and A. V. Suarez. 2006. Multifunctionality and mechanical origins: Ballistic jaw propulsion in trap-jaw ants. *Proceedings of the National Academy of Sciences USA* 103: 12787–92.

Perezgasga, L., J. Q. Jiang, B. Bolival, M. Hiller, E. Benson, M. T. Fuller, and H. White-Cooper. 2004. Regulation of transcription of meiotic cell cycle and terminal differentiation genes by the testis-specific Zn-finger protein matotopetli. *Development* 131(8): 1691–1702.

Pickrell, J. 2003. "Probably the oldest" penis found in spider fossil. Nationalgeographic.com. http://news.nationalgeographic.com/news/2003/10/1006_fossilgenitals.html.

Pliny the Elder. *The Natural History.* Ed. J. Bostock and H. T. Riley. http://www.perseus.tufts.edu/cgi-bin/ptext?lookup=Plin.+Nat.+toc.

Prete, F. 1991. Can females rule the hive? The controversy over honey bee gender roles in British beekeeping texts of the sixteenth–eighteenth centuries. *Journal of the History of Biology* 24: 113–44.

Purchas, Samuel. 1657. *A Theatre of Politicall Flying-Insects.* London: Printed by R. I. for Thomas Parkhurst.

Reed, D. L., J. E. Light, J. M. Allen, and J. J. Kirchman. 2007. Pair of lice lost or parasites regained: The evolutionary history of anthropoid primate lice. *BMC Biology* 5: 7–18.

Rick, F. M., G. C. Rocha, K. Dittmar, C. E. A. Coimbra, K. Reinhard, F. Bouchet, L. F. Ferreira, and A. Araújo. 2002. Crab louse infestation in Pre-Columbian America. *Journal of Parasitology* 88(6): 1266–67.

Ridgel, A. L., R. E. Ritzmann, and P. L. Schaefer. 2003. Effects of aging on behavior and leg kinematics during locomotion in two species of cockroach. *Journal of Experimental Biology* 206: 4453–65.

Roeder, K. D. 1935. An experimental analysis of the sexual behavior of the praying mantis *Mantis religiosa. Biological Bulletin* 49: 203–20.

Rothschild, M. 1985. *British Aposematic Lepidoptera. The Moths and Butterflies of Great Britain and Ireland,* vol. 2. Cossidae—Heliodinidae. Ed. J. Heath and A. M. Emmet. Colchester: Harley Books.

Rutledge, C. 1996. A survey of identified kairomones and synomones used by insect parasitoids to locate and accept their hosts. *Chemoecology* 7: 121–31.

Ryan, C., A. Ghosh, D. Smit, B. Wilson-Boyd, and S. O'Leary. 2006. Adult aural foreign bodies. *The Internet Journal of Otorhinolaryngol-*

ogy 4(2). http://www.ispub.com/ostia/index.php?xmlFilePath
=journals/ijorl/vol4n2/foreign.xml.

Sagan, C. 1995. *The Demon-Haunted World—Science as a Candle in the Dark.*
New York: Random House.

Schofield, C. J. 1977. Sound production in some triatomine bugs. *Physiological Entomology* 2: 43–52.

Siveter, D. J., M. D. Sutton, and D. E. G. Briggs. 2003. An ostracode crustacean with soft parts from the Lower Silurian. *Science* 302: 1749–51.

Sreng, L. 1990. Seducin, male sex pheromone of the cockroach *Nauphoeta cinerea:* Isolation, identification, and bioassay. *Journal of Chemical Ecology* 16: 2899–2912.

Sreng, L., Leoncini, I., and Clement, J. L. 1999. Regulation of sex pheromone production in the male *Nauphoeta cinerea* cockroach: Role of brain extracts, corpora allata (CA), and juvenile hormone (JH). *Archives of Insect Biochemistry & Physiology* 40(4): 165–72.

Staines, G., C. Travis, and T. E. Jayerante. 1973. The queen bee syndrome. *Psychology Today* 7(8): 55–60.

Talbot, F. A. 1912. *Moving Pictures: How They Are Made and Worked.* Philadelphia: J. B. Lippincott. Repr., New York: Arno Press, 1970.

Taylor, J. D. 1978. The earwig: The truth behind the myth. *Rocky Mountain Medical Journal* 75(1): 37–38.

Topsell, E. 1658/1967. *History of Four-Footed Beasts.* London: Da Capo Press.

Trofimov, Y. 2000. As a cheese turns, so turns this tale of many a maggot. *Wall Street Journal,* August 23.

van Engelsdorp, D., R. Underwood, D. Caron, and J. Hayes, Jr. 2007. An estimate of managed colony losses in the winter of 2006–2007: A report commissioned by the Apiary Inspectors of America. *American Bee Journal* 147: 599–603.

Vetter, R. S., and P. K. Visscher. 2004. Daddy-longlegs myth. http://spiders.ucr.edu/daddylonglegs.html.

Wade, N. 2003. The archaeology of maleness reaches back . . . and back again. http://www.nytimes.com/2003/12/07/weekinreview.

Walker, C. 2004. Camel spiders: Behind a sensational e-mail from Iraq. *National Geographic,* June 20.

Wharton, D. R. A. and M. L. Wharton. 1957. The production of sex attractant substance and of oothecae by the normal and irradiated American cockroach, *Periplaneta americana (L.). Journal of Insect Physiology* 1: 229–39.

———. 1959. The effect of radiation on the longevity of the cockroach, *Periplaneta americana*, as affected by dose, age, sex and food intake. *Radiation Research* 11: 600–15.

Wilkin, P. J., and M. H. Williams. 1993. Comparisons of the aerodynamic forces on a flying sphingid moth with those predicted by quasi-steady theory. *Physiological Zoology* 66: 1015–44.

Wiseman, R. 2002. *Queen Bees and Wannabes: Helping Your Daughter Survive Cliques, Gossip, Boyfriends, and Other Realities of Adolescence*. New York: Crown, 2002.

Witchey-Lakshmanan, L. C. 1999. Long-acting control of ectoparasites: A review of collar technologies for companion animals. *Advanced Drug Delivery Reviews* 38: 113–22.

acknowledgments

Many of the chapters in this book were written for the *American Entomologist*, a journal with an appreciative, but, all things considered, small circulation consisting almost entirely of entomologists. I'm very grateful to Alan Kahan and the Entomological Society of America for allowing me to adapt them for a wider audience and for use in this book.

Medieval bestiaries are famous for their extravagant illustrations; indeed, the illustrations were key to convincing readers to accept the reality of the beasts therein. The illustrators faced the unusual challenge of creating images of creatures that they couldn't possibly have ever seen. A modern bestiary requires no less imagination but a different kind of sensibility. I was indeed fortunate that my editor at Harvard, Ann Downer-Hazell, suggested the brilliant and inexhaustively creative Jay Hosler for the task of illustrating these chapters. Jay instantly grasped what I was trying to achieve and succeeded in evoking the style of the medieval bestiary with a twenty-first-century sensibility and sly humor. Ann also deserves heartfelt thanks for her support throughout the entire process of creating this book, ranging from making a suggestion several years ago that ultimately inspired the framework for the book to gently suggesting a more reader-friendly and pronounceable title shortly before going to press. I'm very grateful to Michael Fisher of Harvard University Press, who had faith in me and who fearlessly took a chance on the notion of combining insects and humor, a pairing that on the surface would appear to be about as natural as combining broccoli and ice cream. I also want to thank Kate Brick, a kind and competent copy editor who expertly and tactfully assumed the role of the every(wo)man reader, making sure that entomological references were never too obscure or too ponderous to reach their intended audience.

Nobody can be an expert on all insects—after all, there are more than a million of them—and I'm indebted to my colleagues in the Department of Entomology at the University of Illinois (UIUC) and elsewhere for so generously sharing their expertise with me whenever questions arose. Andy Suarez and Jim Whitfield deserve special note in this regard. Jim, in particular, has the misfortune of having an office across the hall from mine and was often the first place I went with a question. Moreover, he graciously read the entire manuscript, and, with his encyclopedic knowledge of the Class Insecta, patiently pointed out problems, inaccuracies, and my own misperceptions of insect biology. My colleague, office neighbor, and good friend Arthur Zangerl graciously served as an early-stage sounding board and always came unhesitatingly to my aid whenever my meager computer skills were inadequate for coping with a crisis.

I'm in debt as well to the UIUC entomology students. In particular, Martin Hauser, entomological polymath, was my spirit guide to all things both German and six-legged. The general education course I've taught since 1989, Insects and People, is for nonscientists; over the years, students from every conceivable corner of the campus have come to the course (with varying degrees of enthusiasm). Through term papers, projects, and casual conversations, they have helped me keep up with entomological elements of pop culture, including some of the most pervasive insect urban legends.

Off campus, Richard Pollock of the Laboratory of Public Health Entomology at the Harvard School of Public Health shared his exhaustive knowledge of human lice and other human parasites, for which I'm grateful. I also would like to express my gratitude to two anonymous reviewers, who were generous in their praise despite finding an abundance of errors, some of which were typographical but others of which were scientific and would have been excruciatingly embarrassing had they ultimately appeared in print. The book is much improved due to their deep knowledge and diligence; if errors remain, I am entirely to blame for them and proactively embarrassed.

Because these essays often strayed from scientific subjects into cultural realms, I am also indebted to my UIUC colleagues across the campus. Doug Kibbee of the French department and Marianne Kalinke of the German department deserve special mention for service above and beyond the call of duty for help in translating texts and explaining obscure idioms in other languages.

My most important editor was, perhaps surprisingly, a humanist, not a scientist. My husband Richard Leskosky, who in his professional life is a professor of cinema studies, helped in more ways than I can recount. He provided useful references on a staggering diversity of subjects, timely translations of obscure texts in classical languages, careful corrections of grammar and syntax, and gentle reminders that what entomologists find hilarious may not generate the same response in the rest of humanity. This book, along with almost everything else I do, would have been infinitely more difficult without his unstinting support and encouragement. Finally, I thank my daughter Hannah Leskosky, who, despite having no particular interest of her own in insects, has (almost always) happily shared in her mother's peculiar obsession. The butterfly birthday cakes she bakes for me mean more to me than she can possibly imagine.

index

Gorilla lice, 27–28
Grain weevils, 99–100
Grasshoppers, 41, 85, 103
Gromphadorhina portentosa, 20–21
Guardian (U.K.), 81
Gurnee, Ill., 20–21

Habrobracom wasps, 98, 100
Haldane, J. B. S., 105
Halder, G., 49
Hanson, F. B., 98
Hardy, D. E., 45
Harmonia axyridis, 124–125
Harvester butterflies, 72
Harvestman, 144
Hase, A., 22
Hawaii, 45, 71
Hawaiian happyface spiders, 71
Heard, Steve, 132, 134
Helicoverpa zea, 9
Hemiptera, 161
Heterometrus spinifer, 123
Heys, F. M., 98
Hill, David P., 165
Hinton, H. E., 72
Hippocrates, 125
Hiroshima, atomic bombing of, 97, 100
Hocking, Brian, 7
Holland, Philemon, 11
Homoptera, 161
Honey bees, 12, 29–30, 34–36, 71, 113–115, 121–122
Honey dumping, 33
Horridge, G. A., 53
Hottentots, 21

House flies, 39–40
How a Mosquito Operates, 138
Howard, Leland O., 78–80, 91–92
Hurricane Katrina, 163
Hymenoptera, 6, 120

Illinois, 20–21, 61
Independent (U.K.), 32
India, 163
Insanity, 10
Insecta, 52, 129
Insecticides, 18, 59–60, 65, 137
Insects, viii–x, 9, 25, 34, 74; naming of, ix, 10–12, 14, 142, 154–158, 161–162; aerodynamics of, 1–8; in films, 37–40, 45, 51–52, 56, 116, 138, 152; vision of, 38, 41–43; genetically engineered, 44–50; faces on, 70–73; in songs, 80–81; and politics, 86–89; cannibalism in, 90–95; effect of radiation on, 96–101; jumping abilities of, 102–106; and weather prediction, 107–111, 147–150; directional orientation in, 118–123; as aphrodisiacs, 125–129; mutations of, 152–158; zapping of, 164–169
International Herald Tribune, 32
Internet, 15, 18, 21, 26, 32–34, 45, 47, 55, 57–58, 65, 68, 74, 96, 103–104, 118–120, 124–125, 132–133, 137, 142–145, 148–149, 155, 159–163
Iraq, 57–61
Isaacson, Walter, 33